SpringerBriefs in Materials

The SpringerBriefs Series in Materials presents highly relevant, concise monographs on a wide range of topics covering fundamental advances and new applications in the field. Areas of interest include topical information on innovative, structural and functional materials and composites as well as fundamental principles, physical properties, materials theory and design. SpringerBriefs present succinct summaries of cutting-edge research and practical applications across a wide spectrum of fields. Featuring compact volumes of 50 to 125 pages, the series covers a range of content from professional to academic. Typical topics might include

- A timely report of state-of-the art analytical techniques
- A bridge between new research results, as published in journal articles, and a contextual literature review
- A snapshot of a hot or emerging topic
- An in-depth case study or clinical example
- A presentation of core concepts that students must understand in order to make independent contributions

Briefs are characterized by fast, global electronic dissemination, standard publishing contracts, standardized manuscript preparation and formatting guidelines, and expedited production schedules.

More information about this series at http://www.springer.com/series/10111

Huan Pang · Guangxun Zhang ·
Xiao Xiao · Huaiguo Xue

One-dimensional Transition Metal Oxides and Their Analogues for Batteries

Huan Pang
School of Chemistry and Chemical
Engineering
Yangzhou University
Yangzhou, Jiangsu, China

Guangxun Zhang
School of Chemistry and Chemical
Engineering
Yangzhou University
Yangzhou, Jiangsu, China

Xiao Xiao
School of Chemistry and Chemical
Engineering
Yangzhou University
Yangzhou, Jiangsu, China

Huaiguo Xue
School of Chemistry and Chemical
Engineering
Yangzhou University
Yangzhou, Jiangsu, China

ISSN 2192-1091   ISSN 2192-1105   (electronic)
SpringerBriefs in Materials
ISBN 978-981-15-5065-2   ISBN 978-981-15-5066-9   (eBook)
https://doi.org/10.1007/978-981-15-5066-9

This Springer imprint is published by the registered company Springer Nature Singapore Pte Ltd.
The registered company address is: 152 Beach Road, #21-01/04 Gateway East, Singapore 189721, Singapore

Preface

One-dimensional (1D) nanostructures (such as nanowires, nanorods, nanotubes and nanobelts) can shorten the ion diffusion path, increase the electrode/electrolyte contact area, lower the discharge-charge time, offer direct current pathways, accommodate volume expansion, mitigate slow electrochemical kinetics and limit mechanical degradation. Additionally, unlike typical 1D nanostructures, 1D-analogue nanostructures, also known as 1D heteronanostructures, consist of multiple components. The types of 1D-analogue nanostructures include core-shell nanostructures, hollow architectures and other intricate one-dimensional analogue construction. The synergistic effects between each component endow heterostructured electrodes with better electrical conductivity, greater electrochemical cycle stability and reversibility, faster ion transport, improved mechanical stability, etc. As a result, they have been widely applied in different fields for energy storage, catalysis and optics. What is noteworthy is that the promising family of transition metal oxides (TMOs) and composite materials has attracted significant attention due to its attractive properties such as their earth-abundance, high-powered energy storage capabilities and environmental friendliness. Due to these properties, TMOs are promising candidates for future applications in electrochemical energy storage devices.

This book shows an up-to-date overview of the controlled syntheses and battery applications of complex 1D nanomaterials. Furthermore, the strategies for structural upgrading to obtain a high-order architecture based on a 1D morphology, which can include core-shell structures, hollow structures, and their hybrids with various carbon materials, etc., are described in detail. A wide range of practical applications of 1D/1D analogue TMOs as electrode materials for advanced batteries such as lithium/sodium/ potassium/magnesium-ion batteries, Li-S batteries, metal-air batteries are also presented in this book. *One-dimensional Transition Metal Oxides and Their Analogues for Batteries* reviews the recent advances in the architectures, properties, and applications of multifarious 1D nanomaterials for next-generation battery-related applications. It is interesting and useful to a wide readership in various fields of materials science and engineering.

Yangzhou, China   Huan Pang
                                                                       Guangxun Zhang
                                                                            Xiao Xiao
                                                                         Huaiguo Xue

Acknowledgments

This work was supported by the National Natural Science Foundation of China (U1904215, 21671170, 21673203, and 21201010), the Top-notch Academic Programs Project of Jiangsu Higher Education Institutions (TAPP). Program for New Century Excellent Talents of the University in China (NCET-13-0645) and Program for Colleges Natural Science Research in Jiangsu Province (18KJB150036). Postgraduate Research & Practice Innovation Program of Jiangsu Province (XKYCX17-038), the Six Talent Plan (2015-XCL-030), and Qinglan Project of Jiangsu.

Contents

Abbreviations

AC	Activated carbon
AG	Activated graphene
An	Annealed
AR	Array
BIOXs	Biogenous iron oxides
BM	Ball milling
BTS	Bacteria-Templated Synthesis
CA	Calcination
c-b	Core-branch
CBD	Chemical bath-deposition
CC	Carbon Cloth
CFC	Carbon fiber cloth
CFP	Carbon fiber paper
CMFs	Carbon microfibers
CNT	Carbon nanotube
CPP	Cationic polymerization process
c-s	Core-shell
CT	Carbon textiles
CVD	Chemical vapor deposition
CVR	Chemical vapor-reaction
DH	Double hydroxide
DIC	Deposition-immersion-calcination
ECP	Electrodeposition
ED	Electro-deposition
Es	Electrospinning
Gf	Graphite foam
GF	Graphene Foam
Gr	Graphene
H	Hierarchical
H-MT	Hexagonal microtube

Ht	Hydrothermal
Hy	Hydrothermal
IEP	Ion exchange process
LBIOX	Leptothrix ochracea's product
LSR	Liquid solid reaction
MAP	Microwave-assisted process
Mc	Macroporous carbon
MCNTs	Multiwall carbon nanotubes
MEGO	Microwave exfoliated graphite oxide
Mp	Mesoporous
MT	Microtube
MV	Microwave
MWCNT	Multiwall carbon nanotube
NB	Nanobelt
NBrs	Nanobranches
NC	Nanocone
NC	Nanocable
NCs	Nanocables
NFls	Nanoflakes
NFs	Nanofibers
NNs	Nanoneedles
NPs	Nanoparticles
NR	Nanorod
NSs	Nanosheets
NTs	Nanotubes
NWs	Nanowies
OMCNW	Order mesoporous carbon nanowire
OR	Ostwald ripening
P	Porous
PS	Phase separation
PSP	Phase separation process
QDs	Quantum Dots
RFMS	Radio-frequency magnetron sputtering
rGO	Reduced graphene oxide
SA	Self assembly
SM	Staudenmaier method
SPG	Solution-phase growth
SS	Solvothermal synthesis
SSS	"Self-scroll" strategy
T	Templating
tO	Ternary oxide
TO	Thermal oxidation
tp	Template
UGF	Ultrathin graphite foam

VGCF	Vapor grown carbon fiber
VT	Vapor transport
VTM	Vapor transport method
WCE	Wet-chemical etching
y-s	Yolk-shelled

Chapter 1
Synthetic Strategies for One-Dimensional/One-Dimensional Analogue Nanomaterials

Abstract 1D nanostructures have attracted extensive interest in energy storage applications because of their large surface-to-volume ratio and electrode-electrolyte contact area, short ion diffusion distance and charge-discharge time. This chapter shows the various methods for synthesizing 1D/1D analogue nanomaterials, with a special emphasis on new synthetic methodologies, including electrospinning, the Kirkendall effect, Ostwald ripening, heterogeneous contraction, and template-assisted synthesis. These preparation processes are controllable and highly effective for obtaining 1D/1D analogue nanomaterials with different porosities, inner structures, morphologies and combinations.

Keywords One-dimensional architectures · One-dimensional analogue constructions · Transition metal oxides · Nanomaterials · Controllable synthesis

There are two basic methods for synthesizing 1D nanostructures: bottom-up and top-down. A bottom-up process forms the 1D nanostructures by combining constituent adatoms. A top-down process reduces a large block of material to small-sized structures by different means such as lithography and electrophoresis [1–11]. However, most synthesis techniques involve a bottom-up approach. Figure 1.1 shows the architectures, synthesis methods and application of 1D and intricate 1D analogue nanomaterials. Furthermore, Fig. 1.2 shows the crystal structures, different oxidation valences, reaction equations and theoretical capacities of TMOs.

1.1 Thermally Driven Contraction

The porous structure in 1D materials can be obtained by volume or mass loss after thermal decomposition, especially in unbalanced conditions, to generate heterogeneous contraction and produce hollow constructions [9]. The decomposable precursors of simple and complex hollow constructions include MOFs, metal glycerates, carbonates, etc. [12, 13]. There are some similarities among these materials:

© The Author(s), under exclusive license to Springer Nature Singapore Pte Ltd. 2020
H. Pang et al., *One-dimensional Transition Metal Oxides and Their Analogues for Batteries*, SpringerBriefs in Materials, https://doi.org/10.1007/978-981-15-5066-9_1

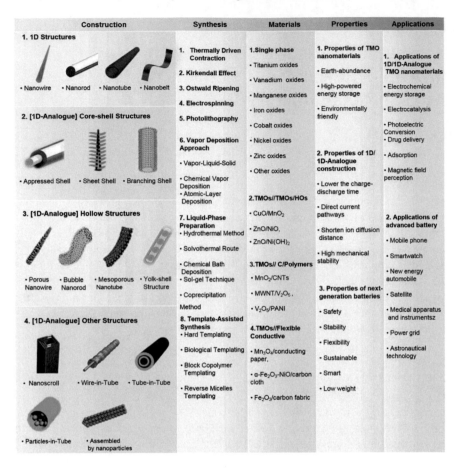

Fig. 1.1 Synthesis methods and architectures of one-dimensional and intricate one-dimensional analogue nanomaterials (in the first two columns). Four typical single-phase and hybrid systems of transition metal oxide-based nanomaterials for batteries (in the third column). Properties and applications of one-dimensional/intricate one-dimensional analogue TMOs and high-performance batteries (in the last two columns)

(1) They consist of a decomposable inorganic/organic moiety and a metallic moiety. Under most circumstances, the metallic portion is decentralized homogeneously in the inorganic/organic portion at the nano or atomic scale.

(2) They usually experience obvious weight loss in the process of thermal decomposition because of the combustion of the organic portion.

(3) For the above materials, material migration could occur in the course of thermal decomposition.

MOFs as functional materials have been widely applied in a variety of fields such as water harvesting from air, [14] energy-efficient dehydration, [15] selective separation, [16–18] catalysts, [19] and so on [20, 21]. At the same time, MOFs are

Titanium Oxides	Vanadium Oxides	Manganese Oxides	Iron Oxides
 Anatase TiO_2	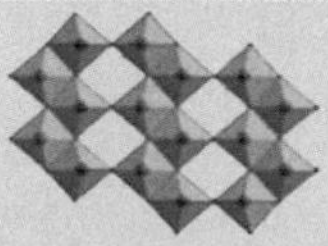 V_2O_5	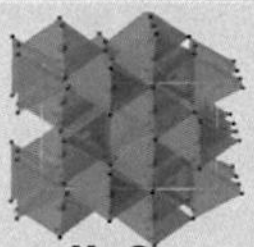 Mn_3O_4	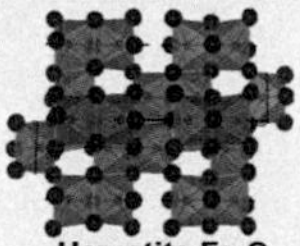 Hematite Fe_2O_3
TiO, Ti_2O_3, Ti_3O_5, Ti_4O_7, Ti_5O_9, Ti_7O_{13}, Ti_9O_{17}...	V_2O_5, VO_2, V_2O_3, V_3O_5, V_3O_7, V_4O_7, V_5O_9, V_6O_{11}, V_6O_{13}...	MnO, Mn_3O_4, Mn_2O_3, MnO_2, MnO_3, Mn_2O_7...	FeO, FeO_2, Fe_3O_4, Fe_4O_5, Fe_5O_6, Fe_5O_7, $Fe_{25}O_{32}$, $Fe_{13}O_{19}$ ...
Triclinic, Rhombohedral, Tetragonal, Cubic, Monoclinic, Orthorhombic phases	Orthorhombic, Cubic, Triclinic, Tetragonal, Monoclinic, Rhombohedral phases	α-MnO_2: alpha phase, β-MnO_2: beta phase γ-MnO_2: gamma phase, ε-MnO_2: epsilon phase	α-Fe_2O_3: alpha phase, β-Fe_2O_3: beta phase γ-Fe_2O_3: gamma phase, ε-Fe_2O_3: epsilon phase
$TiO_2 + Li^+ + e^- = LiTiO_2$	$V_2O_5 + 0.5Li^+ + 0.5e^- = Li_{0.5}V_2O_5$	$Mn_3O_4 + 8Li^+ + 8e^- = 3Mn + 4Li_2O$	$Fe_2O_3 + 6Li_+ + 6e^- = 2Fe + 3Li_2O$
$TiO_2 + Na^+ + e^- = NaTiO_2$	$V_2O_5 + 0.5Na^+ + 0.5e^- = Na_{0.5}V_2O_5$	$Mn_3O_4 + 8Na^+ + 8e^- = 3Mn + 4Na_2O$	$Fe_2O_3 + 6Na^+ + 6e^- = 2Fe + 3Na_2O$
Tc=335 mA h g^{-1}	Tc=294 mA h g^{-1}	Tc=937 mA h g^{-1}	Tc=1007 mA h g^{-1}

Cobalt Oxides	Nickel Oxides	Zinc Oxides	Molybdenum Oxides
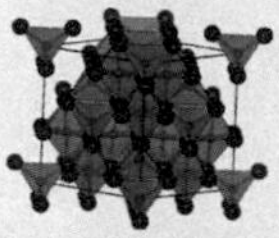 Spinel oxides like Co_3O_4	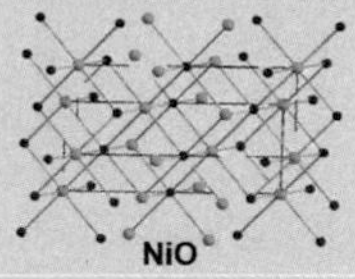 NiO	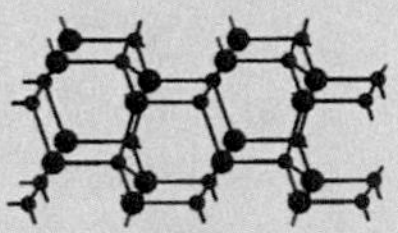 ZnO	 MoO_3
CoO, Co_2O_3, Co_3O_4	NiO, NiO_2, Ni_2O_3	ZnO, ZnO_2	MoO_3, MoO_2, Mo_4O_{11}, Mo_8O_{23}, $Mo_{18}O_{52}$ ...
CoO: Periclase structure, Cubic crystals Co_2O_3: Trigonal crystals Co_3O_4: Spinel structure, Cubic crystals	Cubic, Monoclinic, Rhombohedral, Hexagonal phases	Hexagonal, Cubic phases	Cubic, Orthorhombic, Monoclinic, Triclinic, Hexagonal phases
$Co_3O_4 + 8Li^+ + 8e^- = 3Co + 4Li_2O$	$NiO + 2Li^+ + 2e^- = Ni + Li_2O$	$ZnO + 2Li^+ + 2e^- = Zn + Li_2O$ $Zn + Li^+ + e^- = LiZn$	$MoO_3 + 6Li^+ + 6e^- = Mo + 3Li_2O$
$Co_3O_4 + 8Na^+ + 8e^- = 3Co + 4Na_2O$	$NiO + 2Na^+ + 2e^- = Ni + Na_2O$		$MoO_3 + 6Na^+ + 6e^- = Mo + 3Na_2O$
Tc=890 mA h g^{-1}	Tc=717 mA h g^{-1}	Tc=987 mA h g^{-1}	Tc=1117 mA h g^{-1}

Fig. 1.2 Representative crystal structures, different oxidation valences, reaction equations and theoretical capacities of transition metal oxide electrode materials for advanced batteries

very fascinating precursors for building metal oxide tubular structures because of their rich compositions, easy availability, variety of ligand and metal-ion combinations, and well-defined morphologies [22, 23]. The organic constituent of MOFs will pyrolyze under calcination in air, leaving behind the homologous metal oxides. These MOF-derived 1D metal oxides display enhanced electrochemical properties for Li-ion batteries [24]. For example, Wang and co-workers demonstrated the synthetic strategy of $Fe_2O_3/NiFe_2O_4$ nanotubes by annealing Fe MIL-88/Fe_2Ni MIL-88 MOFs [25]. In the beginning of the calcination process, a thin middle shell with many vacancies appeared on the surface of the composite materials. The vacancies in this shell could allow the outward diffusion of Fe MIL-88/Fe_2Ni MIL-88 and the inward diffusion of oxygen. Because of the faster rate of diffusion of Fe MIL-88/Fe_2Ni MIL-88 compared to atmospheric oxygen, a typical cavity can be obtained. Furthermore, Lou's group reported the synthesis of a yolk–shell Ni/Co oxide nanometer prism using a fast, thermally driven, heterogeneous contraction route [8]. The yolk–shell structure was synthesized with a beneficial porous structure by facile thermal annealing in an air atmosphere. Owing to the coordination ability of PVP for specific metal ions via the C=O and/or –N functional groups, the use of PVP is vital for the formation of discrete and uniform precursor particles (Fig. 1.3a) [8]. Then, an annealing treatment in air is applied to fabricate Ni–Co mixed oxides from the Ni–Co precursor. The formation of yolk–shell architecture is principally based on a heterogeneous contraction mechanism caused by uneven heat-treatment. A TEM image shows the mesoporous yolk–shell feature with a typical void space between the inner core and the outer shell (Fig. 1.3b). The thickness of this shell construction, which was approximately 20–30 nm, was robust enough endure the process of thermal decomposition. In addition, the interior core was highly porous, consisting of plentiful polycrystalline particles. With the excellent porous composition, these Ni–Co oxide yolk–shell nanoprisms showed vastly enhanced electrochemical properties for Li-ion batteries.

1.2 Kirkendall Effect

The Kirkendall effect is the movement of the adjacent boundary layer between two species, and it occurs as a result of the diversity in diffusion rates of the atoms [26]. The Kirkendall effect has a number of important practical applications for the synthesis of porous 1D materials. It can create numerous voids at the boundary interface, and these are referred to as Kirkendall voids. Pores in nanomaterials have ramifications for thermal, electrical and mechanical properties, and thus, the controllable synthesis of voids for 1D nanostructures is often desired. The size of the hole is determined by many factors, including the differential concentration, the intrinsic diffusivities of the materials, the annealing temperature and the annealing time. For instance, a Ag/MnO_x composite nanotubes have been synthesized through the Kirkendall effect under a solvothermal route [27]. The performance of the Ag/MnO_x nanorod was observably enhanced because of the large specific surface area formed

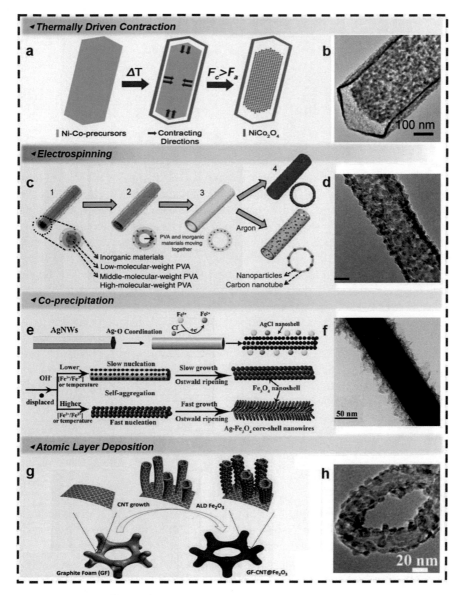

Fig. 1.3 a Schematic of the formation process of Ni/Co oxide nanometer prisms. (Fa: adhesive force and Fc: contraction force). **b** TEM image of yolk–shell $Ni_{0.37}Co$ oxide nanoprisms [8]. Copyright 2015, Wiley-VCH. **c** Schematic illustration of the preparation of mesoporous nanotubes by the gradient electrostatic spinning method and controlled pyrolysis process. **d** TEM image of mesoporous Co_3O_4 nanotubes (the scale bar is 20 nm) [5]. Copyright 2015, Springer Nature. **e** Growth mechanism of Ag nanowires covered by Fe_3O_4 nanoparticles using a co-precipitation process. **f** TEM image of Ag/Fe_3O_4 heterogeneous core–shell nanowires [39]. Copyright 2015, American Chemical Society. **g** Growth mechanism of Fe_2O_3 covered with graphite foam and CNTs. **h** HRTEM image of Fe_2O_3-decorated graphite foam and CNTs [44]. Copyright 2015, American Chemical Society

by using the Kirkendall effect and easier transport by the hierarchical structure. This synthetic strategy and reaction mechanism may provide feasible design guidelines for the preparation of other 1D metal oxide nanomaterials.

1.3 Ostwald Ripening

Ostwald ripening is a representative phenomenon in liquid sols or solid solutions that describes the morphological change of an inhomogeneous structure over time, i.e., sol particles or small crystals dissolve, and redeposit onto sol particles or larger crystals. In general, both the Ostwald ripening and Kirkendall effect have been developed for controllable synthesis of hollow 1D nanostructures. Ostwald ripening methods generally include a two-step process: (1) the synthesis of a template precursor and (2) the transformation of the template into hollow 1D structures. For example, a CuO/MnO_2 core–shell nanostructures has been synthesized using both Ostwald ripening and Kirkendall growth mechanisms without any surfactants [7]. In addition, hierarchical $BaFe_{12}O_{19}$ hollow fibers were successfully fabricated through a convenient Ostwald ripening approach, which used homogeneous electrospinning gel fibers as the precursor. This shell-forming Ostwald ripening route involves a two-step heat-treatment, in which the layered shells produced during the pretreatment process induce outward substance migration to form small crystallites in the initial sections of the hierarchical fibers by the Ostwald ripening mechanism. This is followed by a high-temperature procedure, which results in hollow 1D fibers. At the same time, the in situ-generated Ostwald ripening technique could also be easily extended to manufacture other hollow 1D structures.

1.4 Electrospinning

The electrospinning technique is one of the most extensive and efficient strategies for the synthesis of different 1D nanomaterials [5, 28]. In this process, the precursor material is fed via a spinneret using a syringe pump. At the same time, precursor droplets are deformed and elongated into a Taylor cone under high voltage. Subsequently, a charged jet is continuously ejected from the conical structure and stretched to form nanofibers [29]. Furthermore, electrospinning can be applied to fabricate a more complicated 1D morphology, such as core–shell and porous constructions, which provided significant void space. Currently, various 1D nanotubes, nanowires and intricate nanowires have been fabricated by the electrospinning method. Significantly, the influencing parameters of the electrospinning process include the type of polymers, the ratio of inorganics to polymer and the concentration of the precursor. Additionally, in the preparation process of 1D nanomaterials, annealing plays a significant role in controlling the structure [30]. The influence factor of the annealing process includes the temperature, heating rate, atmosphere, time, etc. Mai and co-workers successfully demonstrated general gradient-electrospinning followed by a controlled pyrolysis process to fabricate multifarious 1D nanostructures, such as continuous nanowires and mesoporous nanotubes, [31] which include binary-metal oxides (LiV_3O_8, $NiCo_2O_4$, $LiCoO_2$ and $LiMn_2O_4$), multi-element oxides ($Na_{0.7}Fe_{0.7}Mn_{0.3}O_2$, $Li_3V_2(PO_4)_3$ and $Na_3V_2(PO_4)_3$) and single-metal oxides (SnO_2, MnO_2, Co_3O_4, and CuO). The critical point of this methodology is the graded distribution of low-, middle- and high-molecular-weight PVA during the electrostatic spinning process [5]. Additionally, these novel mesoporous nanotubes (Fig. 1.3c, d), which present wonderful electrochemical properties in supercapacitors and batteries because of their high conductivity and large specific area will have promising prospects in energy storage and other fields.

1.5 Photolithography

Photolithography, also termed UV lithography or optical lithography, is a technology used in micro-fabrication to pattern parts of a substrate or a thin film. This technique usually uses light to transfer a specific geometric pattern from a photomask to photographic chemistry "photoresist" on the substrate. Subsequently, a suite of chemical treatments then enables either the deposition of or exposure to a pattern to form a new nanomaterial in the requested pattern on the material below the photo resist. This process is a high-precision version of the approach used to fabricate printed circuit boards. Photolithography can achieve extremely diminutive patterns (down to nanometers size), and it provides exact control of both the size and shape of the products it creates. Furthermore, lithographic methods include electron-beam lithography, focused ion beam lithography and dip-pen nanolithography, etc., which can be used to fabricate large-scale nanometer arrays. However, the high cost and low-throughput of large-scale nanomaterials remain a challenge.

1.6 Liquid-Phase Preparation

Liquid-based processes play a significant role in the chemical synthesis of multi-dimensional nanostructures. Based on different reaction conditions including temperature, pressure, concentration, additives time, pH, etc., various nanomaterials have been obtained. At the same time, the liquid-based approach has been continually researched for the manufacture of 1D architectures due to their simple processing, outstanding dispersion, size controllability, low cost and good scalability. However, the crystal construction and growth orientation of 1D materials are not easily controlled via liquid-based methods. This section covers synthetic methods of 1D nanomaterials using liquid-phase technologies, including hydrothermal synthesis, solvothermal route, sol-gel method and the like.

1.6.1 Hydrothermal Method

The hydrothermal route provides a controllable and simple bottom-up method for the synthesis of different dimensional nanomaterials, including 3D, 2D, 1D and 0D structures. Its advantages are confirmed as high yields, a convenient process and energy conservation [32]. At the same time, hydrothermal methods are very efficient for structured 1D architectures, including VO_2, MnO_2, WO_3, [33]. $Ag_2V_4O_{11}$, $Ag_4V_2O_6$, and $Na_2V_6O_{16} \cdot 3H_2O$, etc. For example, Goriparti et al. successfully synthesized carbon-doped TiO_2-bronze NWs by a facile hydrothermal method [34]. This nanostructure was applied as an active nanomaterial for Li-ion batteries. The results demonstrate that the wire geometry and the presence of carbon doping will improve the electrochemical performance of these nanomaterials. Direct carbon doping will improve the electrical conductivity of the nanowires and reduce the lithium-ion diffusion length. Additionally, cycling experiments show that carbon-doped TiO_2-bronze NWs showed a superior rate capability and outstanding higher capacities compared to the undoped nanowires.

1.6.2 Solvothermal Route

The solvothermal route is usually implemented in autoclaves. A supercritical fluid gradually appears with elevated temperature and pressure, which boosts the restricted dissolvability of the predecessor and results in the sediment of representative 1D structure from the excess compound precursor. For example, Chen et al. demonstrated the synthesis of TiO_2 ultrafine nanorods through a one-step solvothermal process utilizing magnesium powder as the reducing agent [35]. Ti^{3+} self-doped TiO_2 nanorods with a diameter of approximately 7 nm exhibits a dark color and was used as an anode material in LIBs for the first time [31, 36]. Furthermore, Ren et al. synthesized an ordered forest construction, which was based on 1D $VO_2(B)$ nanobelts with densely packed vertically aligned structures [37]. They used solvothermal synthesis and a vertical graphene net as a substrate.

1.6.3 Chemical Bath Deposition

Chemical bath deposition (CBD) is an approach to deposit nanomaterials and thin film, it could be employed for continuous or large-scale deposition. The CBD technique is a facile process that could yield adherent, hard, stable and uniform films with good reproducibility. The important parameters of CBD include temperature, concentration, pH, morphology characteristics of the substrate and deposition duration. The reaction mechanism of CBD involves two steps: nucleation and nanoparticle growth. For instance, a series of hierarchically porous TMO arrays and their corresponding mixed nanowire arrays have been reported via chemical bath deposition following a calcination process, and they were grown directly on conductive substrates [38]. When used as conductive-agent-free and binder-free electrodes for LIBs, the resultant nanostructured electrodes manifest excellent electrochemical properties with a superior rate capability, outstanding cycling stability and high specific capacity.

1.6.4 Co-precipitation Method

For example, an $Ag-Fe_3O_4$ core–shell nanowires (Fig. 1.3f) have been manufactured by co-precipitation process [39]. The morphology and size control of the 1D core–shell nanostructure caused by adjusting the reaction conditions such as the time, temperature and concentration of the reactants. The shell thickness of Fe_3O_4 can be adjusted from 76 to 6 nm with various architectures between the nanorods and nanospheres. The growth mechanism of the core–shell nanowires was possibly proposed, as shown in Fig. 1.3e. (1) The C=O of PVP on the Ag nanowire surface offered a nucleation site, and the oxidation reaction between the $FeCl_2/FeCl_3$ solution and the Ag nanowires accelerated the enhancement of Fe^{2+} and Fe^{3+} on the surface of the Ag nanowires. (2) Fe_3O_4 nanoparticles gradually nucleated on the surface of the Ag nanowires. (3) Fe_3O_4 nanoparticles gradually grew on the surface of the Ag nanowires.

1.6.5 Sol–Gel Technique

In the sol-gel technique, the sol is a colloidal suspension manufactured by mixing distilled water with metal organic compounds or metal salts [40]. The sol-gel process has been universally used to fabricate inorganic nanomaterials or inorganic-organic hybrid materials. The sol-gel approach includes the sol of colloidal solutions in the liquid and gel of gelatinized colloidal dissolutions. For example, a $VO_2(B)$ nanobelt forest was synthesized by a solvothermal process using a graphene network as the support [37]. The forest architecture further expanded into a folded 3D forest architecture via graphene-covered metallic foam as the supporting scaffold.

1.7 Vapor Deposition Approach

The vapor deposition approach is generally conducted under vacuum and high temperature. This technique has many advantages. First, the high-temperature and high-vacuum deposition environments allow for high crystallization of 1D metallic oxides. Second, control of the composition, dimension, location and organization of 1D nanocrystals can be obtained using the vapor deposition approach by engineering precursors, catalysts and growth sites.

1.7.1 Atomic Layer Deposition

Atomic layer deposition (ALD) is a self-limited, cycled chemical vapor deposition methodology of different precursors by different precursor pulses. ALD is shown to be very effective for developing nanomaterials of altitudinal uniformity with conformal characteristics, particularly for substrates with a complex morphology and high aspect ratio. This method is universally used to compound isomorphic thin film structures and to control film thicknesses down to the nanoscale. A conformal and thin layer with favorable control of the growth velocity can be deposited via ALD technology, which utilizes continuous self-limiting gas and solid reactions. The typical reaction mechanism of ALD provides numerous advantages, which include accurate control of the thickness, conformal coating and a low-temperature process. ALD has been used to modify electrolyte/electrode interfaces and to manufacture different electrode materials and solid-state electrolytes in the energy storage field. For instance, Ahmed et al. successfully synthesized HfO_2-coated MoO_3 nanorod anodes by ALD [41]. ALD coating on the electrolyte and electrode interface aims to prevent disadvantageous interfacial reactions that occur between the liquid electrolyte and electrode during the discharging and charging processes. Significantly, rationally designed electrode constructions with carbon-based materials are highly desirable and can optimize the overall electrochemical property. Furthermore, the graphite foam carbon nanotubes (Gf/CNTs) can be fabricated from a graphite foam CNT forest using a CVD technique, and they could be utilized as a lightweight and flexible substrate with favorable mechanical stability [42, 43]. As illustrated in Fig. 1.3g, the Gf–CNT composite was made using chemical vapor deposition technology, where a CVD graphite thin foam was used as a substrate [44]. The Gf–CNT exhibited high mechanical flexibility, surface area, and electrical conductivity. With the ALD coating process, a hierarchical architecture of CNT–Gf/Fe_2O_3 was developed. An HRTEM image of the CNT–Gf/Fe_2O_3 construction is shown in Fig. 1.3h. The CNT is congruously covered with typical nanoparticles. The facile strategy of decorating transition metal oxides on lightweight, highly porous, and conductive carbon material supports could be extended to other electrode structures for next-generation, high energy storage.

1.7.2 Chemical Vapor Deposition

Chemical Vapor Deposition (CVD) is a synthetic method in which a functional material reacts with vaporized, decomposed precursors and particular substrate surfaces to form 1D nanomaterials. The CVD processes include low-temperature, microwave plasma, hot filament, plasma-enhanced, and photo assisted CVD. A CVD-derived 3D ultrathin graphite foam is endowed with a wonderful backbone for the preparation of binder-free and high-rate electrodes [45–47]. The CVD-derived 3D ultrathin graphite foam has a porosity of approximately 99.7%, a large specific surface area of approximately 850 $m^2 g^{-1}$, favorable electrical conductivity of approximately 1000 S m^{-1}, and a low density of approximately 0.6 mg cm^{-2}. Based on this method, Chao's group synthesized a V_2O_5/PEDOT nanobelt array grown on an ultrathin graphite foam by a CVD approach [48]. The free-standing, lightweight V_2O_5 nanoarray electrodes were formed by first growing a V_2O_5 nanobelt on 3D ultrathin graphite and then coating the V_2O_5 array with a thin layer of PEDOT. One main component of this route is the free-standing, conductive PEDOT coating layer without the use of surfactant, and the PEDOT shell could promote charge transfer resulting from the reaction kinetics.

1.7.3 Vapor-Liquid-Solid Synthesis

Vapor-liquid-solid (VLS) refers to an aggregation mechanism in which the functional material is first absorbed by liquid catalysts during nanocrystal growth. The VLS route has great prospects for the preparation of single-crystalline nanowires from various inorganic nanomaterials. In the VLS route, the diameter and spatial position of the nanowires could be controlled by adjusting the position and size of the metal catalyst [49]. Additionally, the hetero-nanostructures along the radial or axial directions can be unceasingly designed via the VLS method. Metal oxide nanowires that can be grown using the VLS process are typically MgO, In_2O_3, ZnO, SnO_2, and Ga_2O_3.

1.8 Template-Assisted Synthesis

The template-assisted method is the most generally used technique for the preparation of nanostructures, especially 1D nanomaterials. There are two categories of templates used in manufacturing 1D construction: one is the hard template and the other is the soft template. Hard templates include silica SBA-15, CMK-3, anodic alumina oxide (AAO), and MCM-41, etc. Correspondingly, soft templates include various biological templates, universal block copolymer templates, surfactant templates and so on.

1.8.1 Hard Templating

The hard-templating method is significant for the facile preparation of potential materials with desired functions and controlled nanostructures/microstructures. This technology basically involves four steps: (1) the preparation of a hard template; (2)

surface modification; (3) the introduction of a target product; and (4) removal of the template. Various hard templates with porous or channel structures have been continually researched to form nanorods/tubes. These templates have many characteristics, such as easy shape/size control, easy synthesis, good dispersibility and ready availability in large amounts. For instance, a hierarchical MnO_2 nanofibril/nanowire array (Fig. 1.4a) was rationally designed using an AAO template (Fig. 1.4b). In step one, a MnO_2 nanowire array attached to gold was synthesized by an anodic deposition technique [50]. The next step is an electrochemical reduction process. The result shows that the MnO_2 nanofibrils attached to the surface of the MnO_2 nanowires. This template route provides favorable control over the different parameters for each construction including their length, thickness and diameter. Furthermore, silica SBA-15 was also a suitable hard template for the manufacture of a 1D nanowire construction [51]. Based on this, Mousavi et al. synthesized ordered mesoporous $CuCo_2O_4$ nanowires using an SBA-15 silica template (Supplementary Fig. 1.4c). Figure 1.4d shows a side-view image of the highly ordered $CuCo_2O_4$ materials comprised of nanowires organized into bundles with a length of approximately a few hundred nanometers, resulting from the specific architecture and sufficient electrolyte contact with the active materials using the interconnected channels.

1.8.2   Reverse Micelles Templating

Reverse micelles are identified as shaping vesicles for multifarious nanostructures. They are formed at a decided ratio of water in oil and are considered an isotropic microemulsion. The co-surfactant and surfactant (usually a short chain amine and alcohol) can stabilize the thermodynamic dispersion of the water phase in a consecutive oil phase. The aqueous core of reverse micelles templating acts as a "nanoreactor", which controls the construction of nanomaterials over a narrow size distribution. There are many advantages to the reverse micelles method, such as a homogeneous size distribution, excellent monodispersion, and controlled size. Most importantly, the impact of parameters include the intermicellar exchange rate, W_0 = $[H_2O/surfactant]$, molar ratio of surfactant to water reactant concentration, solvent, etc. [52]. Vaidya et al. varied the proportion of surfactants and made various morphologies such as nanorods, nanocubes and nanospheres and showed that a higher concentration of the solvent could result in larger size of nanorods. Sharma's group synthesized nickel oxalate nanorods via CTAB/isooctane/1-butanol/water microemulsions (with a change in W_0 from 4 to 30) [53]. The experimental results show that the surfactants, the bulk oil phase, and control of the water to surfactant ratio (W_0) are vital for designing the modified conditions needed to make nanomaterials of conceivable shapes and sizes. The bulkiness of the solvent modulates the microemulsions, and the microemulsion droplets encapsulate the structures via influence of the rigidity and curvature of the surfactant thin film and cause a smaller size of the micellar nucleus, which leads to morphologies with a large aspect ratio. The growth of the micellar nucleus and the dynamical interchange in an anisotropic mode

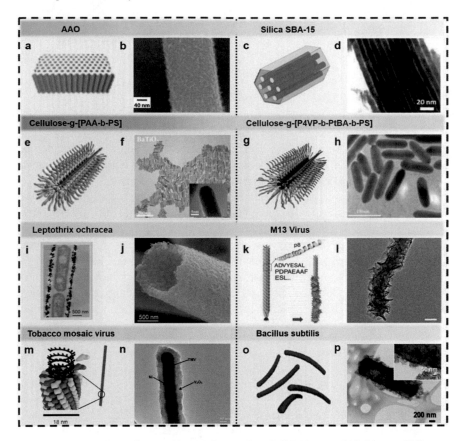

Fig. 1.4 a Anodized aluminum oxide (AAO) template. **b** SEM image of MnO_2 nanofibrils on a MnO_2 nanowire [50]. Copyright 2013, American Chemical Society. **c** Mesoporous silica SBA-15 template. **d** TEM image of the sample from a side view of the $CuCo_2O_4$ nanowire bundle [51]. Copyright 2015, American Chemical Society. **e** The template of cellulose-g-(PAA-b-PS). **f** TEM images of $BaTiO_3$ nanorods. **g** The template of cellulose-g-(P4VP-b-PtBA-b-PS). **h** TEM images of Au–Fe_3O_4 core–shell nanorods [55]. Copyright 2016, Science. **i** TEM of the template of a rod-like bacteria subsection. **j** High-magnification SEM image of Leptothrix ochracea's product [56]. Copyright 2014, American Chemical Society. **k** The M13 virus template and the biotemplating MnO_x composite nanostructure. **l** TEM image of M13 virus-templated bio-MnO_x nanowires with a scale bar of 50 nm [57]. Copyright 2013, Springer Nature. **m** The template of Tobacco mosaic virus. **n** TEM image of the virus-templated TMV/Ni/V_2O_5 composite nanorod [58]. Copyright 2012, American Chemical Society. **o** The template of Bacillus subtilis. **p** TEM image of cobalt oxide and bacteria composite rods [59]. Copyright 2011, American Chemical Society

leads to the progressive formation of 1D nanostructures. Furthermore, surfactant-free vanadium oxide nanobelts were rationally synthesized at atmospheric pressure and room temperature [54]. The key point of this experiment was the use of an organic oxidant that served as the co-surfactant and oxidant during the preparation process and promoted surfactant diversion with a convenient washing protocol. Meanwhile, this facile surfactant removal procedure could be of universal applicability.

1.8.3 Block Copolymer Templating

As a soft-templating method, block copolymer templating (as seen in Fig. 1.4e, g) is a powerful tool for obtaining 1D nanostructures. Lin and co-workers demonstrated the general synthesis of 1D nanocrystals via functional bottlebrush-like block copolymers [55]. Using this method, a variety of plain nanowires, nanotubes and core–shell nanorods were successfully designed and synthesized with well-defined construction by block copolymer templating as nanoreactors (Fig. 1.4f, h). These columnar unimolecular nanoreactors yielded a high degree of control over the composition, anisotropy, architecture and surface chemistry of 1D nanostructures. Central to this effective technique is the rational fabrication of functional block copolymers-composed of a cellulose framework densely grafted with triblock/diblock copolymers of accurately tunable lengths, which serve as nanosized reactors. Moreover, these prepared nanomaterials also include metallic, up-conversion, semiconducting, thermoelectric and ferroelectric 1D nanocrystals. They can serve as models for basic research in the phase behaviors, crystallization kinetics and self-assembly kinetics of 1D nanomaterials.

1.8.4 Biological Templating

Biological templating is also an advanced soft-templating route. In the past five years, the biomimetic and bioinspired preparation of nanomaterials has drawn increasing research. Bionanotechnology is skillfully used to fabricate typical 1D nanostructures via natural bio-assemblies or genetic engineering. For example, new battery materials were successfully designed and synthesized by Leptothrix ochracea (Fig. 1.4i), a common aquatic bacteria, in which silicon, phosphorus and iron are mixed atomically [56]. The biosynthetic tubules exhibited a diameter of approximately 1 μm and an alterable length of up to the centimeter scale, as shown in Fig. 1.4j. This unique structure will be produced because of the oxide nanoparticles precipitate on a biological microtubular template created by organic bacterial excrement. Each tubule is covered with 50–100 nm long and approximately 20 nm wide fibrils, and the inner side is covered with 20–120 nm sized spheres. Leptothrix ochracea's product was fabricated by a washing and drying process and could be used for Li-ion batteries. These works have provided a good direction for the further study of biosynthetic 1D nanomaterials. Furthermore, Belcher et al. reported the facile synthesis of biotemplated manganese oxide nanowires using an aqueous biotemplating reaction, and it could be used as a catalyst in Li–O_2 batteries [57]. These nanowires with a globular surface construction were fabricate by utilizing the M13 virus (Fig. 1.4k). M13 viruses as multipurpose templates have been used to synthesize nanowires with high length-width ratios (such as metal oxides, semiconductors, etc.) with eco-friendly conditions. The M13 virus-mediated manganese oxide nanowires have a diameter of 80 nm. The TEM images are shown in Fig. 1.4l. Additionally, self-assembled V_2O_5/tobacco mosaic virus (TMV) core–shell electrodes (shown in Fig. 1.4n) have been synthesized by virus-templating (Fig. 1.4m) [58]. The core–shell construction was conformally coated on micro-fabricated gold pillar arrays. The active nanomaterial of V_2O_5 was deposited on the entire surface using ALD. Due to the favorable conformality achieved by both the ALD and viral self-assembly methods, the energy density has been promoted by increasing the surface area of the micropillars. Furthermore, thin active coatings of transition metal oxides enable fast charge/discharge rates, simultaneously maintaining high energy densities. Most importantly, with V_2O_5 used as an active material example, the route could be expanded to various metallic oxides that can be deposited by similar deposition techniques. Notably, bacterial templates have enriched the preparation of various micro/nanostructures, especially for mesoporous 1D materials, owing to their unique slender morphology. For instance, porous Co_3O_4 nanotubes were reported using Bacillus subtilis and gram-positive bacteria as soft templates (as seen in Fig. 1.4o) by a high-efficiency biomineralization procedure [59]. The cobalt oxide nanoparticles with a diameter of 2–5 nm were uniformly crystallized on the bacteria surfaces. Additionally, porous, hollow Co_3O_4 nanorods with a surface area of 73.3 m^2 g^{-1} were manufactured through the thermal bacteria removal procedure (Fig. 1.4p). This inexpensive, environmentally benign and facile preparation for TMOs with porous 1D architecture could be used for different fields, including sensors, supercapacitors, batteries and so on.

References

1. Mai L, Tian X, Xu X, Chang L, Xu L (2014) Nanowire electrodes for electrochemical energy storage devices. Chem Rev 114(23):11828–11862. https://doi.org/10.1021/cr500177a
2. Kempa TJ, Day RW, Kim SK, Park HG, Lieber CM (2013) Semiconductor nanowires: a platform for exploring limits and concepts for nano-enabled solar cells. Energ Environ Sci 6(3):719–733
3. Dasgupta NP, Sun J, Liu C, Brittman S, Andrews SC, Lim J, Gao H, Yan R, Yang P (2014) Semiconductor nanowires-synthesis, characterization, and applications. Adv Mater 26(14):2137–2184. https://doi.org/10.1002/adma.201305929
4. Zhang G, Xiao X, Li B, Gu P, Xue H, Pang H (2017) Transition metal oxides with one-dimensional/one-dimensional-analogue nanostructures for advanced supercapacitors. J Mater Chem A 5(18):8155–8186. https://doi.org/10.1039/c7ta02454a
5. Niu C, Meng J, Wang X, Han C, Yan M, Zhao K, Xu X, Ren W, Zhao Y, Xu L, Zhang Q, Zhao D, Mai L (2015) General synthesis of complex nanotubes by gradient electrospinning and controlled pyrolysis. Nat Commun 6:7402
6. Anderson BD, Tracy JB (2014) Nanoparticle conversion chemistry: Kirkendall effect, galvanic exchange, and anion exchange. Nanoscale 6(21):12195–12216. https://doi.org/10.1039/c4nr02025a

7. Huang M, Zhang Y, Li F, Wang Z, Alamusi HuN, Wen Z, Liu Q (2014) Merging of Kirkendall growth and Ostwald ripening: $CuO@MnO_2$ core–shell architectures for asymmetric supercapacitors. Sci Rep 4:4518. https://doi.org/10.1038/srep04518
8. Yu L, Guan B, Xiao W, Lou XW (2015) Formation of yolk–shelled Ni–Co mixed oxide nanoprisms with enhanced electrochemical performance for hybrid supercapacitors and lithium ion batteries. Adv Energy Mater 5(21):1500981. https://doi.org/10.1002/aenm.201500981
9. Yu L, Wu HB, Lou XW (2017) Self-templated formation of hollow structures for electrochemical energy applications. Accounts Chem Res 50(2):293–301. https://doi.org/10.1021/acs.accounts.6b00480
10. Liu Y, Elzatahry AA, Luo W, Lan K, Zhang P, Fan J, Wei Y, Wang C, Deng Y, Zheng G, Zhang F, Tang Y, Mai L, Zhao D (2016) Surfactant-templating strategy for ultrathin mesoporous TiO_2 coating on flexible graphitized carbon supports for high-performance lithium-ion battery. Nano Energy 25:80–90. https://doi.org/10.1016/j.nanoen.2016.04.028
11. Lei D, Benson J, Magasinski A, Berdichevsky G, Yushin G (2017) Transformation of bulk alloys to oxide nanowires. Science 355:267–271
12. Cho W, Lee YH, Lee HJ, Oh M (2011) Multi ball-in-ball hybrid metal oxides. Adv Mater 23(15):1720–1723
13. Shen L, Yu L, Yu X, Zhang X, Lou XW (2015) Self-templated formation of uniform $NiCo_2O_4$ hollow spheres with complex interior structures for lithium-ion batteries and supercapacitors. Angew Chem Int Edit 54(6):1868–1872
14. Kim H, Yang S, Rao SR, Narayanan S, Kapustin EA, Furukawa H, Umans AS, Yaghi OM, Wang EN (2017) Water harvesting from air with metal-organic frameworks powered by natural sunlight. Science 356(6336):430
15. Cadiau A, Belmabkhout Y, Adil K, Bhatt PM, Pillai RS, Shkurenko A, Martineaucorcos C, Maurin G, Eddaoudi M (2017) Hydrolytically stable fluorinated metal-organic frameworks for energy-efficient dehydration. Science 356(6339):731
16. Isfahani AP, Ghalei B, Sivaniah E, Hirao H, Kusuda H, Doitomi K, Wakimoto K, Sakurai K, Song Q, Furukawa S (2017) Enhanced selectivity in mixed matrix membranes for CO_2 capture through efficient dispersion of amine-functionalized MOF nanoparticles. Nat Energy 2:17086
17. Liao PQ, Huang NY, Zhang WX, Zhang JP, Chen XM (2017) Controlling guest conformation for efficient purification of butadiene. Science 356(6343):1193
18. Knebel A, Geppert B, Volgmann K, Kolokolov DI, Stepanov AG, Twiefel J, Heitjans P, Volkmer D, Caro J (2017) Defibrillation of soft porous metal-organic frameworks with electric fields. Science 358(6361):347–351
19. Jagadeesh RV, Murugesan K, Alshammari AS, Neumann H, Pohl MM, Radnik J, Beller M (2017) MOF-derived cobalt nanoparticles catalyze a general synthesis of amines. Science 358(6361):326
20. Zheng S, Li X, Yan B, Hu Q, Xu Y, Xiao X, Xue H, Pang H (2017) Transition-metal (Fe Co, Ni) based metal-organic frameworks for electrochemical energy storage. Adv Energy Mater 7(18):1602733
21. Yu J, Mu C, Yan B, Qin X, Shen C, Xue H, Pang H (2017) Nanoparticle/MOF composites: preparations and applications. Mater Horiz 4(4):557
22. Zhang L, Wu HB, Lou XW (2013) Metal-organic-frameworks-derived general formation of hollow structures with high complexity. J Am Chem Soc 135(29):10664–10672
23. Ju P, Jiang L, Lu TB (2013) An unprecedented dynamic porous metal-organic framework assembled from fivefold interlocked closed nanotubes with selective gas adsorption behaviors. Chem Commun 49(18):1820–1822
24. Hu L, Huang Y, Zhang F, Chen Q (2013) CuO/Cu_2O composite hollow polyhedrons fabricated from metal-organic framework templates for lithium-ion battery anodes with a long cycling life. Nanoscale 5(10):4186
25. Huang G, Zhang F, Zhang L, Du X, Wang J, Wang L (2014) Hierarchical $NiFe_2O_4/Fe_2O_3$ nanotubes derived from metal organic frameworks for superior lithium ion battery anodes. J Mater Chem A 2(21):8048–8053. https://doi.org/10.1039/c4ta00200h

26. Yin Y, Rioux RM, Erdonmez CK, Hughes S, Somorjai GA, Alivisatos AP (2004) Formation of hollow nanocrystals through the nanoscale Kirkendall effect. Science 304(5671):711–714

27. Li Y, Fu H, Zhang Y, Wang Z, Li X (2014) Kirkendall effect induced one-step fabrication of tubular Ag/MnO_x nanocomposites for supercapacitor application. J Phys Chem C 118(13):6604–6611. https://doi.org/10.1021/jp412187n

28. Ren W, Zheng Z, Luo Y, Chen W, Niu C, Zhao K, Yan M, Zhang L, Meng J, Mai L (2015) An electrospun hierarchical LiV_3O_8 nanowire-in-network for high-rate and long-life lithium batteries. J Mater Chem A 3(39):19850–19856. https://doi.org/10.1039/c5ta04643b

29. Wang H, Yuan S, Ma D, Zhang X, Yan J (2015) Electrospun materials for rechargeable batteries: from structure evolution to electrochemical performance. Energ Environ Sci 8(6):1660–1681

30. Peng S, Li L, Hu Y, Srinivasan M, Cheng F, Chen J, Ramakrishna S (2015) Fabrication of spinel one-dimensional architectures by single-spinneret electrospinning for energy storage applications. ACS Nano 9(2):1945

31. Chen X, Liu L, Yu PY, Mao SS (2011) Increasing solar absorption for photocatalysis with black hydrogenated titanium dioxide nanocrystals. Science 331(6018):746–750

32. Shi W, Song S, Zhang H (2013) Hydrothermal synthetic strategies of inorganic semiconducting nanostructures. Chem Soc Rev 42(13):5714

33. Gao L, Wang X, Xie Z, Song W, Wang L, Wu X, Qu F, Chen D, Shen G (2013) High-performance energy-storage devices based on WO_3 nanowire arrays/carbon cloth integrated electrodes. J Mater Chem A 1(24):7167–7173

34. Goriparti S, Miele E, Prato M, Scarpellini A, Marras S, Monaco S, Toma A, Messina GC, Alabastri A, De Angelis F, Manna L, Capiglia C, Zaccaria RP (2015) Direct synthesis of carbon-doped TiO_2-bronze nanowires as anode materials for high performance lithium-ion batteries. ACS Appl Mater Inter 7(45):25139–25146. https://doi.org/10.1021/acsami.5b06426

35. Chen J, Song W, Hou H, Zhang Y, Jing M, Jia X, Ji X (2015) Ti^{3+} self-doped dark rutile TiO_2 ultrafine nanorods with durable high-rate capability for lithium-ion batteries. Adv Funct Mater 25(43):6793–6801. https://doi.org/10.1002/adfm.201502978

36. Chen X, Liu L, Huang F (2015) Black titanium dioxide (TiO_2) nanomaterials. Chem Soc Rev 44(7):1861

37. Ren G, Hoque MNF, Pan X, Warzywoda J, Fan Z (2015) Vertically aligned $VO_2(B)$ nanobelt forest and its three-dimensional structure on oriented graphene for energy storage. J Mater Chem A 3(20):10787–10794. https://doi.org/10.1039/c5ta01900a

38. Zhang Q, Wang J, Dong J, Ding F, Li X, Zhang B, Yang S, Zhang K (2015) Facile general strategy toward hierarchical mesoporous transition metal oxides arrays on three-dimensional macroporous foam with superior lithium storage properties. Nano Energy 13:77–91. https://doi.org/10.1016/j.nanoen.2015.01.029

39. Ma J, Wang K, Zhan M (2015) Growth mechanism and electrical and magnetic properties of Ag–Fe(3)O(4) core–shell nanowires. ACS Appl Mater Inter 7(29):16027–16039. https://doi.org/10.1021/acsami.5b04342

40. Ma FX, Yu L, Xu CY, Lou XW (2016) Self-supported formation of hierarchical $NiCo_2O_4$ tetragonal microtubes with enhanced electrochemical properties. Energ Environ Sci 9(3):862–866. https://doi.org/10.1039/c5ee03772g

41. Ahmed B, Shahid M, Nagaraju DH, Anjum DH, Hedhili MN, Alshareef HN (2015) Surface passivation of MoO_3 nanorods by atomic layer deposition toward high rate durable li ion battery anodes. ACS Appl Mater Inter 7(24):13154–13163. https://doi.org/10.1021/acsami.5b03395

42. Liu J, Zhang L, Wu HB, Lin J, Shen Z, Lou XW (2014) High-performance flexible asymmetric supercapacitors based on a new graphene foam/carbon nanotube hybrid film. Energ Environ Sci 7(11):3709–3719

43. Liu J, Chen M, Zhang L, Jiang J, Yan J, Huang Y, Lin J, Fan HJ, Shen ZX (2014) A flexible alkaline rechargeable Ni/Fe battery based on graphene foam/carbon nanotubes hybrid film. Nano Lett 14(12):7180–7187

44. Guan C, Liu J, Wang Y, Mao L, Fan Z, Shen Z, Zhang H, Wang J (2015) Iron oxide-decorated carbon for supercapacitor anodes with ultrahigh energy density and outstanding cycling stability. ACS Nano 9:5198

45. Ji H, Zhang L, Pettes MT, Li H, Chen S, Shi L, Piner R, Ruoff RS (2012) Ultrathin graphite foam: a three-dimensional conductive network for battery electrodes. Nano Lett 12(5):2446
46. Luo J, Liu J, Zeng Z, Ng CF, Ma L, Zhang H, Lin J, Shen Z, Fan HJ (2013) Three-dimensional graphene foam supported Fe_3O_4 lithium battery anodes with long cycle life and high rate capability. Nano Lett 13(12):6136
47. Yu X, Lu B, Xu Z (2014) Super long-life supercapacitors based on the construction of nanohoneycomb-like strongly coupled CoMoO(4)-3D graphene hybrid electrodes. Adv Mater 26(7):1044–1051
48. Chao D, Xia X, Liu J, Fan Z, Ng CF, Lin J, Zhang H, Shen ZX, Fan HJ (2014) A V_2O_5/conductive-polymer core/shell nanobelt array on three-dimensional graphite foam: a high-rate, ultrastable, and freestanding cathode for lithium-ion batteries. Adv Mater 26(33):5794–5800. https://doi.org/10.1002/adma.201400719
49. Klamchuen A, Suzuki M, Nagashima K, Yoshida H, Kanai M, Zhuge F, He Y, Meng G, Kai S, Takeda S, Kawai T, Yanagida T (2015) Rational concept for designing vapor-liquid-solid growth of single crystalline metal oxide nanowires. Nano Lett 15(10):6406–6412. https://doi.org/10.1021/acs.nanolett.5b01604
50. Duay J, Sherrill SA, Gui Z, Gillette E, Lee SB (2013) Self-limiting electrodeposition of hierarchical MnO_2 and $M(OH)_2$/MnO_2 nanofibril/nanowires: mechanism and supercapacitor properties. ACS Nano 7(2):1200
51. Pendashteh A, Moosavifard SE, Rahmanifar MS, Wang Y, ElKady MF, Kaner RB, Mousavi MF (2015) Highly ordered mesoporous $CuCo_2O_4$ nanowires, a promising solution for high-performance supercapacitors. Chem Mater 27(11):3919–3926. https://doi.org/10.1021/acs.chemmater.5b00706
52. Sharma S, Pal N, Chowdhury PK, Sen S, Ganguli AK (2012) Understanding growth kinetics of nanorods in microemulsion: a combined fluorescence correlation spectroscopy, dynamic light scattering, and electron microscopy study. J Am Ceram Soc 134(48):19677–19684
53. Sharma S, Yadav N, Chowdhury PK, Ganguli AK (2015) Controlling the microstructure of reverse micelles and their templating effect on shaping nanostructures. J Phys Chem B 119(34):11295–11306. https://doi.org/10.1021/acs.jpcb.5b03063
54. Tartaj P, Amarilla JM, Vazquez-Santos MB (2015) Surfactant-free vanadium oxides from reverse micelles and organic oxidants: solution processable nanoribbons with potential applicability as battery insertion electrodes assembled in different configurations. Langmuir ACS J Surf Colloids 31(45):12489–12496. https://doi.org/10.1021/acs.langmuir.5b02856
55. Pang X, He Y, Jung J, Lin Z (2016) 1D nanocrystals with precisely controlled dimensions, compositions, and architectures. Science 353(6305):1268
56. Hashimoto H, Kobayashi G, Sakuma R, Fujii T, Hayashi N, Suzuki T, Kanno R, Takano M, Takada J (2014) Bacterial nanometric amorphous Fe-based oxide: a potential lithium-ion battery anode material. ACS Appl Mater Inter 6(8):5374–5378. https://doi.org/10.1021/am500905y
57. Oh D, Qi J, Lu YC, Zhang Y, Shao-Horn Y, Belcher AM (2013) Biologically enhanced cathode design for improved capacity and cycle life for lithium-oxygen batteries. Nat Commun 4:2756. https://doi.org/10.1038/ncomms3756
58. Gerasopoulos K, Pomerantseva E, Mccarthy M, Brown A, Wang C, Culver J, Ghodssi R (2012) Hierarchical three-dimensional microbattery electrodes combining bottom-up self-assembly and top-down micromachining. ACS Nano 6(7):6422
59. Shim HW, Jin YH, Seo SD, Lee SH, Kim DW (2011) Highly reversible lithium storage in bacillus subtilis-directed porous Co_3O_4 nanostructures. ACS Nano 5:443

Chapter 2
Daedal Construction for One-Dimensional/One-Dimensional Analogue Nanomaterials

Abstract It is a general trend that 1D nanomaterials constructed with multiple functionalities and novel morphologies are researched for enhanced performance and applications. As mentioned in this chapter, 1D/1D analogue nanomaterials can be fabricated by liquid-phase and gaseous-phase techniques among other techniques, and they are promising candidates for energy storage devices including supercapacitors and batteries. Apart from this, various 1D analogue architecture, such as core–shell, hollow and other intricate 1D analogue nanostructures, usually have more robust electrochemical properties than the primary 1D construction. Therefore, synthetic strategies for their improvement need to be investigated and explored. These novel, intricate, 1D analogue structures will be reviewed in this chapter.

Keywords One-dimensional architectures · One-dimensional analogue construction · Core–shell morphologies · Hollow morphologies

2.1 Core–Shell One-Dimensional Analogue Structures

The core–shell nanostructure is an effective architecture to synergistically enhance both the properties and materials, [1] and it has been widely used in various fields, including catalysis, photovoltaics, [2–5] water splitting, [6] nanoelectronics and biotechnology, etc. Especially in energy storage equipment, core–shell structures could improve the chemical stability, prevent structural changes, buffer the stress of the inner material, enhance the electrical conductivity, and prevent degradation and aggregation [7–14]. In recent years, various materials, including conducting polymers, [15] metal hydroxides/oxides [16] and carbons, [17, 18] have been fabricated with this type of core–shell construction. In practical applications, one structural model is not always enough, and a complex core–shell construction system could be represented by a range of representative models, from single-level to high-level models. In this article, we provide step-by-step examples showing ways to fabricate various core–shell structures, including coaxial, layered and branching core–shell nanostructures.

2.1.1 Core–Shell Structure with Cable-Like Morphology

A pressed core–shell construction is a correspondingly elementary core–shell structure, and it usually consists of two mutually nested tubular nanostructures. A low
conductivity and the structural degradation of transition metal oxides could result in
capacity fading in lithium batteries. Coating or combining another nanosized material (such as a conductive additive) on the 1D backbone is the most common and
facile approach to synthesizing cable-like nanostructures. For instance, Amine et al.
synthesized CuO/CN_x core–shell nanostructures with cable-like morphologies, and
they could be used for large-area and flexible anode assemblies [19]. These free-
standing 3D nanoarrays with core–shell architectures, which could be fabricated
by the oxidative growth of CuO nanowires onto Cu substrates and radio-frequency
sputtering of CN_x films. A theoretical architectural model and TEM image of core–
shell CuO/CN_x nanocables could be seen in Fig. 2.1a, b, respectively. The cable-like
CuO/CN_x core–shell construction could adapt to the volume change during the process of lithiation and delithiation, and simultaneously, the 3D arrays could offer
ample ion/electron transport paths and electroactive zones. In addition, the mono-
lithic cable-like nanostructure without additional conductive agents or binders could
enhance the power density and energy density of the whole electrode. Additionally, novel transition metal oxide/polymer core–shell nanowires with cable-like morphologies have be synthesized, and they were grown on conductive, lightweight, thin,
porous graphite foams [20]. The fabrication process of the core/shell $Co_3O_4/PEDOT$–
MnO_2 nanowire is schematically illustrated in Fig. 2.1c. As shown in Fig. 2.1d, this
cable-like nanowire has a Co_3O_4 nanowire core and a MnO_2–PEDOT composite
shell.

2.1.2 Core–Shell Structures with Layered Shell

A MoS_2 nanosheets attached to TiO_2 nanobelts have been successfully synthesized
via a simple hydrothermal method [25]. The TiO_2/MoS_2 delivers promising lithium-
ion storage properties with good rate performances, high specific capacities and
stable cyclability. In addition, a hierarchically core–shell nanostructures have been
rational designed by a hydrothermal process [21]. The schematic in Fig. 2.1e shows
the fabrication process of the TiO_2 and $H_2Ti_3O_7$ nanowires with a MnO_2-nanoflake
core–shell construction. TEM images of the MnO_2/TiO_2 hierarchical nanowire are
shown Fig. 2.1f. The construction has a shell of mesoporous MnO_2 sheets and a
core of TiO_2–B nanowires. Significantly, the layered double hydroxides are promising candidates for electrochemical energy conversion and storage, because of their
environmentally friendly nature, tunable chemical composition and cost effective-
ness [26–29]. Duan et al. designed and synthesized core–shell CuO/layered double hydroxide nanostructures on a thin Cu wire substrate by a three-step procedure
(Fig. 2.1g): [22] (1) $Cu(OH)_2$ nanowire arrays were grown on a Cu wire; (2) $Cu(OH)_2$

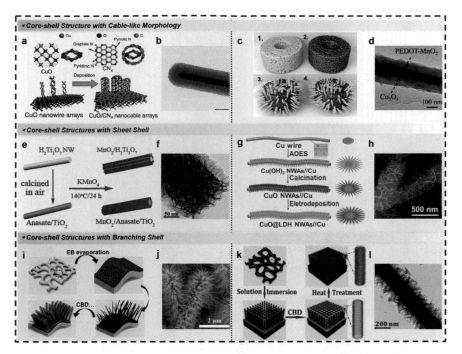

Fig. 2.1 a Theoretical architectural model showing the formation mechanism of CuO nanowires and CuO/CN$_x$ core–shell nanocables. **b** TEM image of a CuO/CN$_x$ core–shell nanocable (the scale bars are 200 nm) [19]. Copyright 2016, Springer Nature. **c** The fabrication process of Co$_3$O$_4$/PEDOT–MnO$_2$ core/shell nanowires arrays on porous graphite foams. (1) 3D porous Ni films. (2) 3D graphite foams. (3) Co$_3$O$_4$ nanowires. (4) Co$_3$O$_4$/PEDOT–MnO$_2$ nanowire arrays. **d** TEM image of a Co$_3$O$_4$/PEDOT–MnO$_2$ core/shell nanowire [20]. Copyright 2014, American Chemical Society. **e** Schematic diagram of the processes of manufacturing MnO$_2$ nanoflakes on TiO$_2$ (B) and H$_2$Ti$_3$O$_7$ nanowires. **f** TEM image of the MnO$_2$/TiO$_2$ hierarchical nanowires [21]. Copyright 2014, Elsevier. **g** Diagrammatic drawing for the manufacture of core/shell CuO/CoFe–LDH nanowire arrays on a Cu wire. **h** SEM image of hierarchical CuO/CoFe-LDH nanowire arrays [22]. Copyright 2016, Elsevier. **i** Schematic illustration for the manufacturing procedure of hierarchical CuO/Co$_3$O$_4$ core/shell nanowire arrays on Ni foam. **j** SEM images of CuO/Co$_3$O$_4$ core/shell hetero-architectured nanowires [23]. Copyright 2014, Elsevier. **k** Manufacturing process of 3D CuO/CoO core/shell arrays with hierarchical tubular construction on Cu foam. **l** TEM image of a hierarchical CuO/CoO core/shell architecture [24]. Copyright 2015, Elsevier

was translated into the CuO nanowire arrays via heat-treatment; and (3) The shell of layered double hydroxides nanoflakes was synthesized on the surface of CuO nanowire arrays. Finally, core–shell CuO/CoFe-layered double hydroxide nanostructures (Fig. 2.1h) were obtained by an electrosynthetic process. The layered double hydroxides nanoflakes possessed a thickness of approximately 12 nm and a lateral size of approximately 150 nm.

2.1.3   Core–Shell Structures with Branched Shell

Zhang and co-workers synthesized hierarchical Co_3O_4/CuO nanoarrays composed of hierarchical Co_3O_4 branches shell and a CuO nanowire core on Ni foam (Supplementary Fig. 2.1j) using a thermal oxidation process combined with CBD technology followed by a calcination method (as seen in Fig. 2.1i) [23]. These hierarchical core–shell arrays could serve as conductive-agent-free and binder-free anodes for Li-ion batteries, which would enhance the electrochemical properties with a good rate capability and excellent cycle performance compared to single Co_3O_4 nanosheets or pure CuO nanowires. Wang et al. rationally designed and synthesized tubular CuO/CoO core/shell nanoarrays by a controllable fabrication technique [24]. The schematic in Fig. 2.1k shows the fabrication process of core/shell CuO/CoO arrays, which form three-dimensional tubular core/shell heterogeneous nanostructure arrays. TEM image of core/shell CuO/CoO hierarchical architectures could be seen in Fig. 2.1l. The experiment demonstrates that the core/shell architecture could be used for lithium-ion batteries, which exhibit enhanced cycling capabilities and an enhanced capacity.

2.2   Hollow One-Dimensional Analogue Structures

1D hollow nanostructures, which have large surface-to-volume ratios and surface areas, afford convenient transport pathways for ions and electrons [30, 31]. Hollow 1D nanoarchitectures, which combine the advantages of hollow structures and 1D nanomaterials, have had a vital influence in the field of energy storage devices. Hollow and porous nanomaterials with a high structural stability and large surface area have been widely used in various fields, such as gas storage, energy storage, adsorption, sensing, separation, catalysis, and other areas [32, 33]. Numerous reports demonstrate that creating porous and hollow 1D structures leads to distinct improvements in the electrochemical properties [33–46]. This chapter introduces numerous technologies, such as the template method, hydrothermal methods and electrospinning, to synthesize core–shell nanostructures.

2.2.1   Porous Nanowires

Qiao et al. designed hybrid multi-hole nanowire arrays composed of carbon and Co_3O_4 by the carbonization of an MOF grown on Cu foil, as shown in Fig. 2.2a [47]. A single nanowire with a diameter of approximately 250 nm is observed in the TEM image (Fig. 2.2b). The porous Co_3O_4/C nanowires possess a carbon content of 52.1 wt.% and a surface area of 251 $m^2\ g^{-1}$, which first could be used as the electrode for the oxygen evolution reaction without employing additional binders

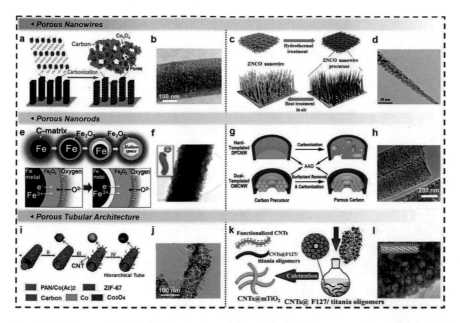

Fig. 2.2 **a** Scheme of the formation process of porous Co_3O_4/carbon nanowire arrays. **b** TEM image of hybrid Co_3O_4/C nanowires [47]. Copyright 2014, American Chemical Society. **c** Manufacturing process of the mesoporous Zn/Ni/Co ternary oxide nanowire arrays. **d** TEM image of individual mesoporous Zn/Ni/Co ternary oxide nanowires [48]. Copyright 2015, American Chemical Society. **e** Formation process of a Fe_2O_3 sphere in a Fe_2O_3/C bubble–nanorod-architectured nanofiber by the Kirkendall effect (above) and the process of chemical conversion in the surface of a nanosphere (below). **f** TEM image of the Fe_2O_3/C bubble–nanorod-architectured nanofibers and its structure schematic drawing (left top) [49]. Copyright 2015, American Chemical Society. **g** Schematic representation of various mesoporous material fabricated by DPCNW and OMCNW templates. **h** TEM image of OMCNW/Fe_2O_3 composites [50]. Copyright 2016, Royal Society of Chemistry. **i** Scheme of the manufacturing process of the hierarchical CNT/Co_3O_4 tubular construction. **j** TEM image of the CNT/Co_3O_4 tubular construction [51]. Copyright 2016, Wiley-VCH. **k** Schematic illustrations of the fabrication process of mesoporous CNTs/TiO_2 nanocables using a surfactant template. **l** TEM images of a mesoporous CNTs/$mTiO_2$ nanocable, and the inset is a structure model [52]. Copyright 2016, Elsevier

or substrates. The composite electrode achieved a remarkably strong durability and oxygen evolution activity with enhanced performance compared to nonmetal and transition-metal/noble-metal catalysts due to the in situ carbon incorporation and specific nanowire array electrode configuration, which generate enhanced mass-charge transport capabilities, a strong structural stability, an easy release of oxygen gas bubbles and a large active surface area. Furthermore, Zhang et al. synthesized mesoporous Zn/Ni/Co ternary oxide nanowire arrays by a hydrothermal and calcination process [48]. The schematic in Fig. 2.2c shows the formation process of porous Zn/Ni/Co ternary oxide nanowires. (Figure 2.2d) Moreover, the diameter of each nanowire was approximately 10–30 nm, and the length of the nanowire was approximately 2–3 μm. These highly porous 1D nanostructures were composed of nanometer crystals of approximately 5 nm in size. The typical mesoporous morphology is shown to be favorable for rapid electron/ion transfer and electrolyte penetration, thereby resulting in an enhanced electrochemical reaction.

2.2.2 Porous Nanorods

A "bubble–nanorod composite" has been synthesized by electrospinning with the Kirkendall effect [49]. The Fe_2O_3/C composite bubble–nanorod nanofibers were synthesized from hollow Fe_2O_3 nanospheres dispersed uniformly in the amorphous carbon matrix. Subsequently, the FeO_x/carbon nanofibers composite nanostructure under an air atmosphere at 300 °C produces the bubble–Fe_2O_3/C nanorod composite nanofibers (as seen in Fig. 2.2f). At the same time, Fig. 2.2e illustrates the formation mechanism and chemical conversion process of hollow Fe_2O_3 nanospheres. The reduction of FeO_x produced the solid Fe nanocrystals that were then converted into a hollow Fe_2O_3 nanosphere structure during further heating by the famous Kirkendall diffusion process. The volume change of the hollow nanospheres is accommodated during cycling. The structure of the bubble–Fe_2O_3/C nanorod composite nanofibers has contributed to enhanced electrochemical properties by enhancing the structural stability during long-term cycling. Moreover, ordered mesoporous carbons (OMCs) are a promising substance that could afford the favorable ion accessibility and desirable electrical conductivity [50]. It establishes the correlations among the electrochemical performance, pore structure and material morphology. Based on this rationale, Hu et al. synthesized an ordered mesoporous carbon nanowires (OMCNWs)/Fe_2O_3 hybrid nanostructure utilizing an especially hard-soft dual-template process. The electrochemical performance and structure of OMCNW/Fe_2O_3 (Fig. 2.2h) were compared with carbon nanowires/Fe_2O_3 and OMCs/Fe_2O_3 composites. The disordered porous carbon nanowire (DPCNW) and OMCNW exhibit bubble-like pores and arch-like shape, which are stable enough to resist the volume change (shown in Fig. 2.2g). The resulting ordered mesoporous carbon nanowires/Fe_2O_3 hybrid nanostructure enables enhanced electrical conductivity, a high structural stability, enhanced Li^+ accessibility and a high pore volume.

2.2.3 Porous Tubular Architectures

Porous and hierarchical tubular nanostructures have many advantages including enlarged pore volume and contact area of electrolyte/electrode. Furthermore, hybrid nanostructures of carbon-based species and inorganic nanostructures for lithium battery applications have attracted great attention recently. There is a popular belief that these hybrid materials could improve both electrical conductivity and mechanical stability in single inorganic materials [53–56]. For example, Lou et al. synthesized hierarchical, tubular, carbon nanotubes and hollow Co_3O_4 nanoparticles [51]. Starting from $Co(Ac)_2$ nanofibers, and subsequently, uniform ZIF-67 core–shell construction were synthesized by partial phase transformation. After the selective dissolution process, the resulting ZIF-67 nanotubes could be converted into Co–C/carbon nanotube composites by annealing. In the end, hierarchical CNT/Co_3O_4 microtubes were manufactured via thermal treatment. The schematic in Fig. 2.2i shows the manufacturing process of the CNT/Co_3O_4 hierarchical microtubes. The CNT/Co_3O_4 hierarchical microtubes could be seen in Fig. 2.2j. Moreover, Zhao et al. realized the assembly of an ultrathin porous TiO_2 shell on flexible graphitized carbon by a surfactant-templating of the amphiphilic triblock copolymer Pluronic F127 [52]. Figure 2.2k illustrates the manufacturing process of the CNTs/TiO_2 porous nanocables. The CNTs/TiO_2 core–shell porous nanotubes (Fig. 2.2l) show an internal pore volume of approximately 0.26 cm^3 g^{-1}, a super-high surface area of approximately 137 m^2 g^{-1}, ultrathin mesoporous TiO_2 shells of approximately 20 nm in thickness and homogeneous mesopores of approximately 6.2 nm. This surfactant-template guided coating approach could be easily extended to deposit a porous TiO_2 layer on planar graphene to fabricate sandwich-like flexible graphene/TiO_2 nanostructures, which provide a method for depositing ultrathin mesoporous oxides on graphitized carbon for energy storage and conversion, including sensing, photocatalysis and so on.

2.2.4 Peapod-Like Structures

TMOs are favorable anode materials with great promise for a variety of batteries because of their high capacities. Nevertheless, these anodes have shortcomings such as low electrical conductivities and obvious volume changes during the Li^+ transfer process. To address the problem, peapod-like Co_3O_4/CNT arrays has been successfully designed. The synthesis process is shown in Fig. 2.3a [40]. The Co_3O_4 nanoparticles are confined in the intracanal air holes of the tubular nanoarrays. Figure 2.3b shows a TEM image of a peapod-like Co_3O_4/CNT architecture. The air holes in the nanotubes are open, and therefore, the electrolyte could availably diffuse onto the surface of the Co_3O_4 nanoparticles. Furthermore, the CNTs could serve as a favorable conductive network.

Additionally, Mai et al. designed a gradient electrostatic spinning and controlled pyrolysis technology to synthesize the pea-like nanotubes architecture, as shown in Fig. 2.3c [45]. As shown in Fig. 2.3d, the $Na_{0.7}Fe_{0.7}Mn_{0.3}O_2$ peapod-like nanotubes were synthesized through the following mechanism:

(1) After the process of electrostatic spinning, the nanowires were immediately and directly placed into a heated furnace in air, which was maintained at 300 °C.
(2) The PVA decomposed and moved to the outer PVA layer without carrying the inorganic functional materials, leaving them in the central zone.

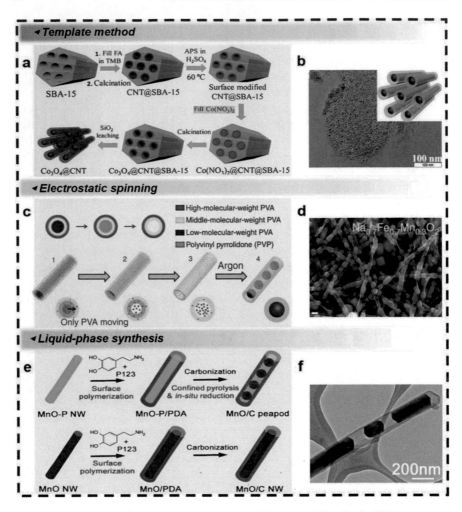

Fig. 2.3 a Schematics of the manufacturing process of the peapod-like Co_3O_4/CNT mesoporous nanotube. (TMB: 1,2,4-trimethylbenzene). **b** TEM image of Co_3O_4/CNT peapod-like construction. The insets is the corresponding illustration [40]. Copyright 2015, Wiley-VCH. **c** Schematic illustration of the gradient electrostatic spinning and controlled pyrolysis technology. **d** TEM image of peapod-like $Na_{0.7}Fe_{0.7}Mn_{0.3}O_2$ nanotubes with a scale bars at 200 nm [45]. Copyright 2015, Springer Nature. **e** Schematic images of the preparation of the pea-like MnO/C heterostructured and MnO/C core–shell nanowires. **f** TEM image of the pea-like MnO/C heterostructured [57]. Copyright 2014, American Chemical Society

(3) After sintering, the outer PVA carbonized, and at the same times, the inner inorganic functional materials converted to nanoparticles, gradually forming pea-like nanotubes.

Owing to the volumetric expansion of MnO nanowires or nanoparticles during lithiation, long-life MnO-based materials for high-performance lithium-ion batteries is still a challenge. Therefore, Jiang et al. synthesized MnO/C with a peapod-like nanostructure by a facile synthetic method [57]. First, the MnO precursor nanowires and polydopamine core/shell structure was prepared by annealing in an inert gas. This synthetic method is different from the synthesis of the typical MnO/C core/shell nanowire nanostructure due to the annealing process with the MnO nanowires-carbon precursor nanostructure (as seen in Fig. 2.3e). The peapod-like MnO/C nanostructure with internal void spaces (Fig. 2.3f) can address the issues related to volumetric expansion during lithium-ion insertion/extraction, conversion, aggregation and MnO dissolution, which improves the electrochemical performance of lithium batteries.

2.3 Other Intricate One-Dimensional Analogue Construction

Multilevel micro/nanomaterials with significantly enhanced electrochemical performance are considered desirable candidates for advanced batteries and supercapacitors due to the synergistic properties and novel architectures. To conquer the limitations of simple-structured electrode materials, multilevel micro/nanomaterials have been a focus as promising nanostructures in the energy storage field because they share many advantages including large surface areas, high body/surface ratios, more surface active sites and better permeability.

2.3.1 Hierarchical Nanoscrolls

Hierarchical nanoscrolls with tunable spaces provide relative interlayer sliding to adjust for volume changes, which can increase the stability of batteries. Mai et al. first synthesized vanadium oxide hierarchical nanoscrolls using Ostwald ripening and scroll-by-scroll and self-rolling processes. The schematic is illustrated in Fig. 2.4a [58]. The polygonal nanoscrolls exist in various shapes, including pentagon, triangle, quadrangle and so on, and the morphological feature of the nanoscrolls is spiral-wrapped multiwalled (as seen in Fig. 2.4b, c), which is beneficial for improving the rate capability and cycling stability in electrochemical energy storage applications. This example also has great potential and will be interesting for other related fields. A semi-hollow graphene scroll architecture was fabricated by a nanowire template and constructed through "self-scroll" and "oriented assembly" methods. These semi-hollow graphene scrolls could exhibit lengths of over 30 micron with internal cavities between the graphene scrolls and the nanowires [59]. This unique semi-hollow structure resulted in excellent cycling performance and energy storage capacities because they provide sufficient ion and electron transfer space and channels for the volume expansion of 1D nanomaterials during cycling. This understanding and strategy could

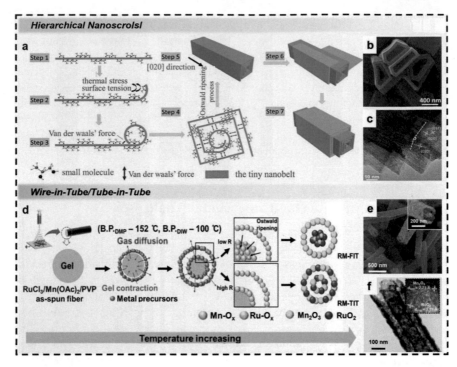

Fig. 2.4 a Mechanism illustration on the preparation of the vanadium oxide nanoscrolls, which includes the processes of Ostwald ripening, self-roll, and scroll-by-scroll. **b** FESEM image of the as-synthesized polygonal nanoscrolls. **c** TEM image of an end of the nanoscrolls [58]. Copyright 2015, Wiley-VCH. **d** Schematic of the manufacturing process of RuO_2/Mn_2O_3 tube-in-tube (RM-TIT) and fiber-in-tube (RM-FIT). **e** SEM image of RuO_2/Mn_2O_3 fiber-in-tube architecture. **f** HRTEM image of RuO_2/Mn_2O_3 tube-in-tube architecture [61]. Copyright 2016, American Chemical Society

be used to synthesize other 1D templated graphene scroll constructions, which can be applied to other fields and fabrication processes.

2.3.2 Wire-in-Tubes/Tube-in-Tubes

Metal oxide constructions with multideck interiors have great potential for energy storage. Metal oxide multideck nanotubes with adjustable interior structures have been reported using electrospinning technology followed by a heat-treatment process [60]. Consequently, this ingenious route may push the continued development of 1D nanostructures. Two types of RuO_2 and Mn_2O_3 composite fibers, namely, multicomposite tube-in-tube and phase-separated fiber-in-tube architectures, have been synthesized by electrospinning technology [61]. The schematic illustration is shown in Fig. 2.4d. These double-walled RuO_2/Mn_2O_3 composite fibers could be fabricated

by controlling the ramping speed during the process of electrostatic spinning. Moreover, both hollow RuO_2/Mn_2O_3 tube-in-tube (Fig. 2.4f) and fiber-in-tube (Fig. 2.4e)
constructions showed remarkable electrochemical properties for the OER and ORR
in $Li–O_2$ cells.

2.3.3   Abundant Particles in Nanotubes

Nanostructured materials have many advantages, such as short distances for ion
transmission, large specific surface areas, and so on [62, 63]. Nevertheless, there
are still many challenges with nanostructured electrodes, such as a high interparticle
resistance, increased surface energy and easy self-aggregation, which lead to degradation of the specific capacity and mechanical stability [64]. In addition, this effect
could accelerate side reactions with the electrolyte and reduce the coulombic efficiency, with a sequence of accumulated SEI layers finally impeding Li-ion transport
[65]. To solve this problem, Liang et al. manufactured CuO particles in Cu nanotube
structures as anode materials [66]. Different from the typical peapod-like structure,
the particles in the nanotube construction are abundant in nanoparticles. This design
provides many attractive features:

(1)   The formation of stable SEI layers and the availability of space for the volumetric
      expansion of the CuO nanoparticles;
(2)   The characteristic one-dimensional analogue nanostructure with a large length-
      width ratio to reduce the process of self-aggregation;
(3)   A nanoscale copper shell to minimize the ionic and electronic impedance.

2.3.4   One-Dimensional Structures Assembled
by Nanoparticles

A 1D array structure assembled from functional Fe_3O_4 nanoparticles has been successfully designed by the capillary-bridge-mediated assembly process [67]. This
route permits the cleavage of a continuous aqueous film of magnetite and the formation of distinct microscale capillary bridges with high aspect ratios and a 1D geometry.
The stacking of magnetite nanospheres could reach several millimeters long with the
evaporation of water and could form sub-micrometer wide 1D arrays with super-
high aspect ratios of more than 5000 with strict precise positioning and alignment.
Furthermore, magnetite Fe_3O_4 nanoparticles were selected because of their good
dispersibility in water, uniform size, and room temperature superparamagnetism.
Consequently, bioinspired 1D Fe_3O_4 nanoparticles arrays have been synthesized via
a columnar structure template with asymmetric wettability. The magnetic anisotropy
could be controlled via changing the widths of the magnetic material to alter the
aspect ratio of the 1D morphology. This also provides inspiration: the bioinspired

magnetic perception technique not only promotes the essential understanding and real operation of anisotropic superparamagnetic nanomaterials but also offers an effective and facile route for the preparation of high-aspect-ratio 1D structures for other fields, including subminiature electronic equipment, biometric techniques, and so on.

References

1. Tan C, Chen J, Wu XJ, Zhang H (2018) Epitaxial growth of hybrid nanostructures. Nat Rev Mater 3:17089
2. Kim SK, Day RW, Cahoon JF, Kempa TJ, Song KD, Park HG, Lieber CM (2012) Tuning light absorption in core/shell silicon nanowire photovoltaic devices through morphological design. Nano Lett 12(9):4971
3. Tang J, Huo Z, Brittman S, Gao H, Yang P (2011) Solution-processed core–shell nanowires for efficient photovoltaic cells. Nat Nanotechnol 6(9):568–572
4. Hwang YJ, Wu CH, Hahn C, Jeong HE, Yang P (2012) Si/InGaN core/shell hierarchical nanowire arrays and their photoelectrochemical properties. Nano Lett 12(3):1678–1682
5. Zhang F, Ding Y, Zhang Y, Zhang X, Wang ZL (2012) Piezo-phototronic effect enhanced visible and ultraviolet photodetection using a ZnO–CdS core–shell micro/nanowire. ACS Nano 6(10):9229
6. Cheng C, Ren W, Zhang H (2014) 3D TiO_2/SnO_2 hierarchically branched nanowires on transparent FTO substrate as photoanode for efficient water splitting. Nano Energy 5(1):132–138
7. Ghosh CR, Paria S (2012) Core/shell nanoparticles: classes, properties, synthesis mechanisms, characterization, and applications. Chem Rev 112(4):2373
8. Xia X, Zhang Y, Fan Z, Chao D, Xiong Q, Tu J, Zhang H, Fan HJ (2015) Novel metal@carbon spheres core–shell arrays by controlled self-assembly of carbon nanospheres: a stable and flexible supercapacitor electrode. Adv Energy Mater 5(6):1401709
9. Zhang G, Wang T, Yu X, Zhang H, Duan H, Lu B (2013) Nanoforest of hierarchical Co_3O_4@$NiCo_2O_4$ nanowire arrays for high-performance supercapacitors. Nano Energy 2(5):586–594
10. Zhao MQ, Zhang Q, Huang JQ, Wei F (2012) Hierarchical nanocomposites derived from nanocarbons and layered double hydroxides-properties, synthesis, and applications. Adv Funct Mater 22(4):675–694
11. Wang J, Zhang X, Wei Q, Lv H, Tian Y, Tong Z, Liu X, Hao J, Qu H, Zhao J, Li Y, Mai L (2016) 3D self-supported nanopine forest-like Co_3O_4 @$CoMoO_4$ core–shell architectures for high-energy solid state supercapacitors. Nano Energy 19:222–233. https://doi.org/10.1016/j.nanoen.2015.10.036
12. Zhao J, Chen J, Xu S, Shao M, Zhang Q, Wei F, Ma J, Wei M, Evans DG, Xue D (2014) Hierarchical Ni Mn layered double hydroxide/carbon nanotubes architecture with superb energy density for flexible supercapacitors. Adv Funct Mater 24(20):2921
13. Xia X, Tu J, Zhang Y, Wang X, Gu C, Zhao XB, Fan HJ (2012) High-quality metal oxide core/shell nanowire arrays on conductive substrates for electrochemical energy storage. ACS Nano 6(6):5531
14. Xia X, Chao D, Qi X, Xiong Q, Zhang Y, Tu J, Zhang H, Fan HJ (2013) Controllable growth of conducting polymers shell for constructing high-quality organic/inorganic core/shell nanostructures and their optical-electrochemical properties. Nano Lett 13(9):4562–4568. https://doi.org/10.1021/nl402741j
15. Han L, Tang P, Zhang L (2014) Hierarchical Co_3O_4 @PPy@MnO_2 core–shell–shell nanowire arrays for enhanced electrochemical energy storage. Nano Energy 7(7):42–51

16. He YB, Li GR, Wang ZL, Su CY, Tong YX (2011) Single-crystal ZnO nanorod/amorphous and nanoporous metal oxide shell composites: Controllable electrochemical synthesis and enhanced supercapacitor performances. Energ Environ Sci 4(4):1288–1292

17. Xiao X, Peng X, Jin H, Li T, Zhang C, Gao B, Hu B, Huo K, Zhou J (2013) Freestanding mesoporous VN/CNT hybrid electrodes for flexible all-solid-state supercapacitors. Adv Mater 25(36):5091

18. Liao JY, Higgins D, Lui G, Chabot V, Xiao X, Chen Z (2013) Multifunctional TiO_2– C/MnO_2 Core–Double-Shell nanowire arrays as high-performance 3D electrodes for lithium ion batteries. Nano Lett 13(11):5467

19. Tan G, Wu F, Yuan Y, Chen R, Zhao T, Yao Y, Qian J, Liu J, Ye Y, Shahbazian-Yassar R (2016) Freestanding three-dimensional core–shell nanoarrays for lithium-ion battery anodes. Nat Commun 7:11774

20. Xia X, Chao D, Fan Z, Guan C, Cao X, Zhang H, Fan HJ (2012) A new type of porous graphite foams and their integrated composites with oxide/polymer core/shell nanowires for supercapacitors: structural design, fabrication, and full supercapacitor demonstrations. Nano Lett 14(3):1651–1658

21. Zhang YX, Kuang M, Hao XD, Liu Y, Huang M, Guo XL, Yan J, Han GQ, Li J (2014) Rational design of hierarchically porous birnessite-type manganese dioxides nanosheets on different one-dimensional titania-based nanowires for high performance supercapacitors. J Power Sources 270(3):675–683

22. Li Z, Shao M, Zhou L, Zhang R, Zhang C, Han J, Wei M, Evans DG, Duan X (2016) A flexible all-solid-state micro-supercapacitor based on hierarchical CuO@layered double hydroxide core–shell nanoarrays. Nano Energy 20:294–304

23. Wang J, Zhang Q, Li X, Xu D, Wang Z, Guo H, Zhang K (2014) Three-dimensional hierarchical Co_3O_4/CuO nanowire heterostructure arrays on nickel foam for high-performance lithium ion batteries. Nano Energy 6:19–26. https://doi.org/10.1016/j.nanoen.2014.02.012

24. Wang J, Zhang Q, Li X, Zhang B, Mai L, Zhang K (2015) Smart construction of three-dimensional hierarchical tubular transition metal oxide core/shell heterostructures with high-capacity and long-cycle-life lithium storage. Nano Energy 12:437–446. https://doi.org/10.1016/j.nanoen.2015.01.003

25. Mao M, Mei L, Guo D, Wu L, Zhang D, Li Q, Wang T (2014) High electrochemical performance based on the TiO_2 nanobelt@few-layered MoS_2 structure for lithium-ion batteries. Nanoscale 6(21):12350–12353. https://doi.org/10.1039/c4nr03991b

26. Shao M, Zhang R, Li Z, Wei M, Evans DG, Duan X (2015) ChemInform abstract: layered double hydroxides toward electrochemical energy storage and conversion: design, synthesis and applications. Chem Commun 51(88):15880

27. Tang C, Wang HS, Wang HF, Zhang Q, Tian GL, Nie JQ, Wei F (2015) Spatially confined hybridization of nanometer-sized nife hydroxides into nitrogen–doped graphene frameworks leading to superior oxygen evolution reactivity. Adv Mater 27(30):4516–4522

28. Wang Q, O'Hare D (2012) Recent advances in the synthesis and application of layered double hydroxide (LDH) nanosheets. Chem Rev 112(7):4124–4155

29. Shao M, Li Z, Zhang R, Ning F, Wei M, Evans DG, Duan X (2015) Hierarchical conducting polymer@clay core–shell arrays for flexible all-solid-state supercapacitor devices. Small 11(29):3530

30. Yu L, Hu H, Wu HB, Lou XW (2017) Complex hollow nanostructures: synthesis and energy-related applications. Adv Mater 29(15):1604563. https://doi.org/10.1002/adma.201604563

31. Wei Q, Xiong F, Tan S, Huang L, Lan EH, Dunn B, Mai L (2017) Porous one-dimensional nanomaterials: design, fabrication and applications in electrochemical energy storage. Adv Mater 29(20):1602300. https://doi.org/10.1002/adma.201602300

32. Perego C, Millini R (2013) Porous materials in catalysis: challenges for mesoporous materials. Chem Soc Rev 42(9):3956

33. Kong B, Selomulya C, Zheng G, Zhao D (2015) New faces of porous Prussian blue: interfacial assembly of integrated hetero-structures for sensing applications. Chem Soc Rev 44(22):7997

34. Qin K, Kang J, Li J, Shi C, Li Y, Qiao Z, Zhao N (2015) Free-standing porous carbon nanofiber/ultrathin graphite hybrid for flexible solid-state supercapacitors. ACS Nano 9(1):481–487
35. Ji L, Gu M, Shao Y, Li X, Engelhard MH, Arey BW, Wang W, Nie Z, Xiao J, Wang C (2014) Controlling SEI formation on SnSb–porous carbon nanofibers for improved Na ion storage. Adv Mater 26(18):2901–2908
36. Aravindan V, Lee Y, Madhavi S (2015) Research progress on negative electrodes for practical Li-ion batteries: beyond carbonaceous anodes. Adv Energy Mater 5(13):1402225
37. Yu XY, Yu L, Lou XW (2016) Metal sulfide hollow nanostructures for electrochemical energy storage. Adv Energy Mater 6(3):1501333
38. Zhang L, Zhang G, Wu HB, Yu L, Lou XW (2013) Hierarchical tubular structures constructed by carbon-coated SnO(2) nanoplates for highly reversible lithium storage. Adv Mater 25(18):2589–2593
39. Zhao Y, Wu W, Li J, Xu Z, Guan L (2014) Encapsulating MWNTs into hollow porous carbon nanotubes: a tube-in-tube carbon nanostructure for high-performance lithium ulfur batteries. Adv Mater 26(30):5113
40. Gu D, Li W, Wang F, Bongard H, Spliethoff B, Schmidt W, Weidenthaler C, Xia Y, Zhao D, Schuth F (2015) Controllable synthesis of mesoporous peapod-like Co_3O_4@carbon nanotube arrays for high-performance lithium-ion batteries. Angew Chem Int Edit 54(24):7060–7064. https://doi.org/10.1002/anie.201501475
41. Li Z, Zhang J, Lou XW (2015) Hollow carbon nanofibers filled with MnO_2 nanosheets as efficient sulfur hosts for lithium–sulfur batteries. Angew Chem Int Edit 54(44):12886–12890. https://doi.org/10.1002/anie.201506972, https://doi.org/10.1002/ange.201506972
42. Wei Q, An Q, Chen D, Mai L, Chen S, Zhao Y, Hercule KM, Xu L, Minhas-Khan A, Zhang Q (2014) One-Pot synthesized bicontinuous hierarchical $Li_3V_2(PO_4)$3/C mesoporous nanowires for high-rate and ultralong-life lithium-ion batteries. Nano Lett 14(2):1042
43. An Q, Lv F, Liu Q, Han C, Zhao K, Sheng J, Wei Q, Yan M, Mai L (2014) Amorphous vanadium oxide matrixes supporting hierarchical porous Fe_3O_4/graphene nanowires as a high-rate lithium storage anode. Nano Lett 14(11):6250–6256. https://doi.org/10.1021/nl5025694
44. Li Q, Wang ZL, Li GR, Guo R, Ding LX, Tong YX (2012) Design and synthesis of MnO_2/Mn/MnO_2 sandwich-structured nanotube arrays with high supercapacitive performance for electrochemical energy storage. Nano Lett 12(7):3803
45. Niu C, Meng J, Wang X, Han C, Yan M, Zhao K, Xu X, Ren W, Zhao Y, Xu L, Zhang Q, Zhao D, Mai L (2015) General synthesis of complex nanotubes by gradient electrospinning and controlled pyrolysis. Nat Commun 6:7402
46. Li Z, Zhang JT, Chen YM, Li J, Lou XW (2015) Pie-like electrode design for high-energy density lithium–sulfur batteries. Nat Commun 6:8850
47. Ma TY, Dai S, Jaroniec M, Qiao SZ (2014) Metal-organic framework derived hybrid Co_3O_4–carbon porous nanowire arrays as reversible oxygen evolution electrodes. J Am Chem Soc 136(39):13925–13931. https://doi.org/10.1021/ja5082553
48. Wu C, Cai J, Zhang Q, Zhou X, Zhu Y, Shen PK, Zhang K (2015) Hierarchical mesoporous zinc–nickel–cobalt ternary oxide nanowire arrays on nickel foam as high-performance electrodes for supercapacitors. ACS Appl Mater Inter 7(48):26512–26521. https://doi.org/10.1021/acsami.5b07607
49. Cho JS, Hong YJ, Kang YC (2015) Design and synthesis of bubble–nanorod-structured Fe_2O_3–carbon nanofibers as advanced anode material for li-ion batteries. ACS Nano 9(4):4026–4035
50. Hu J, Sun CF, Gillette E, Gui Z, Wang Y, Lee SB (2016) Dual-template ordered mesoporous carbon/Fe_2O_3 nanowires as lithium-ion battery anodes. Nanoscale 8(26):12958–12969. https://doi.org/10.1039/c6nr02576e
51. Chen YM, Yu L, Lou XW (2016) Hierarchical tubular structures composed of Co_3O_4 hollow nanoparticles and carbon nanotubes for lithium storage. Angew Chem Int Edit 128(20):6094–6097. https://doi.org/10.1002/anie.201600133, https://doi.org/10.1002/ange.201600133
52. Liu Y, Elzatahry AA, Luo W, Lan K, Zhang P, Fan J, Wei Y, Wang C, Deng Y, Zheng G, Zhang F, Tang Y, Mai L, Zhao D (2016) Surfactant-templating strategy for ultrathin mesoporous TiO_2

coating on flexible graphitized carbon supports for high-performance lithium-ion battery. Nano Energy 25:80–90. https://doi.org/10.1016/j.nanoen.2016.04.028

53. Huang X, Cui S, Chang J, Hallac PB, Fell CR, Luo Y, Metz B, Jiang J, Hurley PT, Chen J (2015) A hierarchical tin/carbon composite as an anode for lithium-ion batteries with a long cycle life. Angew Chem Int Edit 54(5):1490–1493

54. Yu X, Hu H, Wang Y, Chen H, Lou XW (2015) Ultrathin MoS_2 nanosheets supported on N-doped carbon nanoboxes with enhanced lithium storage and electrocatalytic properties. Angew Chem Int Edit 54(25):7395–7398

55. Huang G, Zhang F, Du X, Qin Y, Yin D, Wang L (2015) Metal organic frameworks route to in situ insertion of multiwalled carbon nanotubes in Co_3O_4 polyhedra as anode materials for lithium-ion batteries. ACS Nano 9(2):1592–1599

56. Wang H, Dai H (2013) Strongly coupled inorganic-nano-carbon hybrid materials for energy storage. Chem Soc Rev 42(7):3088–3113

57. Jiang H, Hu Y, Guo S, Yan C, Lee PS, Li C (2014) Rational design of MnO/carbon nanopeapods with internal void space for high-rate and long-life Li-ion batteries. ACS Nano 8(6):6038–6046

58. Wei Q, Tan S, Liu X, Yan M, Wang F, Li Q, An Q, Sun R, Zhao K, Wu H, Mai L (2015) Novel polygonal vanadium oxide nanoscrolls as stable cathode for lithium storage. Adv Funct Mater 25(12):1773–1779. https://doi.org/10.1002/adfm.201404311

59. Yan M, Wang F, Han C, Ma X, Xu X, An Q, Xu L, Niu C, Zhao Y, Tian X, Hu P, Wu H, Mai L (2013) Nanowire templated semihollow bicontinuous graphene scrolls: designed construction, mechanism, and enhanced energy storage performance. J Am Chem Soc 135(48):18176–18182. https://doi.org/10.1021/ja409027s

60. Meng J, Niu C, Liu X, Liu Z, Chen H, Wang X, Li J, Chen W, Guo X, Mai L (2016) Interface-modulated approach toward multilevel metal oxide nanotubes for lithium-ion batteries and oxygen reduction reaction. Nano Res 9(8):2445–2457. https://doi.org/10.1007/s12274-016-1130-x

61. Yoon KR, Lee GY, Jung JW, Kim NH, Kim SO, Kim ID (2016) One-dimensional RuO_2/Mn_2O_3 hollow architectures as efficient bifunctional catalysts for lithium–oxygen batteries. Nano Lett 16(3):2076–2083. https://doi.org/10.1021/acs.nanolett.6b00185

62. Duan X, Huang H, Xiao S, Deng J, Zhou G, Li Q, Wang T (2016) 3D hierarchical CuO mesocrystals from ionic liquid precursors: towards better electrochemical performance for Li-ion batteries. J Mater Chem A 4(21):8402–8411

63. Zhang H, Zhang G, Li Z, Qu K, Wang L, Zeng W, Zhang Q, Duan H (2016) Ultra-uniform CuO/Cu in nitrogen-doped carbon nanofibers as a stable anode for Li-ion batteries. J Mater Chem A 4(27):10585–10592

64. Zhang H, Yu X, Braun PV (2011) Three-dimensional bicontinuous ultrafast-charge and -discharge bulk battery electrodes. Nat Nanotechnol 6(5):277–281

65. Wu H, Yu G, Pan L, Liu N, Mcdowell MT, Bao Z, Cui Y (2013) Stable Li-ion battery anodes by in-situ polymerization of conducting hydrogel to conformally coat silicon nanoparticles. Nat Commun 4(3):1943

66. Zhao Y, Mu S, Sun W, Liu Q, Li Y, Yan Z, Huo Z, Liang W (2016) Growth of copper oxide nanocrystals in metallic nanotubes for high performance battery anodes. Nanoscale 8(48):19994

67. Jiang X, Feng J, Huang L, Wu Y, Su B, Yang W, Mai L, Jiang L (2016) Bioinspired 1D superparamagnetic magnetite arrays with magnetic field perception. Adv Mater 28(32):6952–6958. https://doi.org/10.1002/adma.201601609

Chapter 3
Brief Overview of Next-Generation Batteries

Abstract Due to their flexibility and electrochemical performance, the development of a variety of advanced batteries is experiencing rapid growth for applications in electric vehicles and smart portable devices. It is clear that the product specifications of various electrode materials are being expanded, and this has become a hot research topic in recent years. In this chapter, a systematic review is given of the use of 1D TMOs as electrode materials for various types of batteries, such as Na-ion batteries, Li-ion batteries, K-ion batteries, Li–S batteries, redox flow batteries, metal-air batteries, and hybrid energy storage devices.

Keywords Li-ion batteries · Na-ion batteries · Li–S batteries · K-ion batteries · Redox flow batteries · Metal-air batteries · Hybrid energy storage devices

Transition metal oxides are widely studied as next-generation batteries materials because of their low cost and high specific capacity of nearly three times that of ordinary graphite anodes [1–11]. Figure 3.1 illustrates the trends towards sustainability for today's battery systems. Non-aqueous rechargeable batteries occupy top of the Fig. 3.1a and correspond to the high-voltage area, while the aqueous rechargeable batteries lie at the bottom of the diagram correspond to the lower-voltage area. Simultaneously, Fig. 3.2 shows the Structure comparison diagram of different advanced batteries (Table 3.1).

3.1 Lithium-Ion Batteries

Lithium-ion batteries have led the marketplace of energy storage for use in a variety of electronics since their successful commercialization in the 1990s [84]. With the development of technology, lithium-ion batteries have also been studied extensively for smart grids and electric vehicles [85–87]. A typical lithium-ion battery cell consists of an anode and a cathode, and the two electrodes are connected by an electrolyte and are separated via an ion-permeable membrane [88, 89]. The discharging and charging processes involve Faradaic reactions, which include mass and charge

© The Author(s), under exclusive license to Springer Nature Singapore Pte Ltd. 2020   35
H. Pang et al., *One-dimensional Transition Metal Oxides and Their Analogues for Batteries*, SpringerBriefs in Materials, https://doi.org/10.1007/978-981-15-5066-9_3

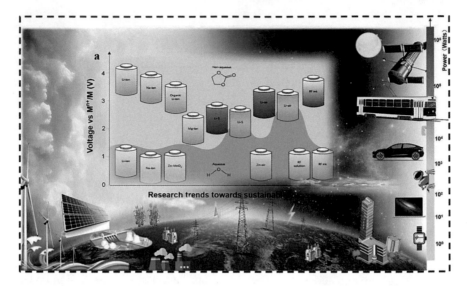

Fig. 3.1 Clean renewable energy system and application of batteries. **a** Trends towards sustainability for today's batteries [12]. Copyright 2017, Springer Nature

transport within the electrodes [4]. Consequently, the performance improvement of functional electrode materials plays a critical role in the batteries [90–94]. Commercial graphite anode materials provide a Li storage capacity via the intercalation of Li ions, and they exhibit a low theoretical capacity of 372 mAh g^{-1}. Functional nanomaterials that store Li$^+$ ions via alloying or conversion reactions could provide much higher relative capacities, and they have been widely research as alternatives to traditional graphite anodes [95–97].

3.2 Lithium–Sulfur Batteries

Among the various contenders in the field of energy storage beyond lithium, lithium–sulfur batteries have appeared as expressly promising because of their potential to reversibly store electrical energy with a low cost and a high theoretical energy density [98]. Nevertheless, there are many challenges associated with Li–S battery systems, such as self-discharge, low sulfur loading, a low active material capacity factor, poor cycling stability and the low electroconductibility of sulfur, which observably restrict its widespread and practical application. The most troublesome problems are the "shuttle effect" between the anode and cathode and the cycling stability to trap the solvable polysulfides in the cathodic area [99–102]. One feasible strategy to solve these obstacles and increase the usability of Li–S batteries is to restrict polysulfides and sulfur in multi-hole conductive materials, for example, conductive polymer or carbon matrices (graphene, [103–105] porous carbon, [106] and carbon nanotubes)

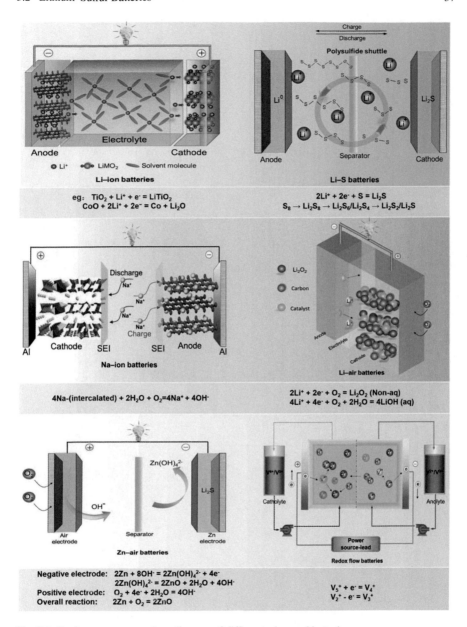

Fig. 3.2 Performance comparison diagram of different advanced batteries

Table 3.1 Composition, architecture, preparation technology and electrochemical performance of 1D/1D Analogue nanomaterials in lithium ion batteries (Cp = Capacity/mAh g⁻¹, Cd = Current densities, Pr = Percentage of retention, Cd = Current densities)

Materials	Synthesis method	Reversible Cp		Rate capacity		Capacity retention			References
		Cp	Cd	Cp	Cd	Pr%	Cycle	Cd	
C/TiO$_2$–B NWs	Hy	306	0.1 C	160	10 C	81	1000	10 C	[13]
Cu/TiO$_2$–B NWs	MV/Hy	187	10 C	150	60 C	64	2000	10 C	[14]
TiO$_2$ NRs	SS	186	0.2 C	123	10 C	98	1000	50 C	[15]
CNTs/TiO$_2$ NCs	Surfactant-T	320	0.4 C	210	20 C	–	–	–	[16]
SnO$_2$/TiO$_2$ c-b ARs	ALD/Hy	580	1.6 A g⁻¹	–	–	–	–	–	[17]
TiO$_2$ NB/MoS$_2$	Hy	710	0.1 A g⁻¹	417	1 A g⁻¹	98	100	0.1 A g⁻¹	[18]
V$_2$O$_5$/PEDOT c-s NB	SS/ED	265	5 C	168	60 C	98	1000	60 C	[19]
V$_2$O$_5$ NWs/VO$_x$ NTs	Hy/An	223	5 A g⁻¹	115	0.5 A g⁻¹	–	–	–	[20]
VO$_2$(B) NB AR/Gr	SS	475	0.1 A g⁻¹	100	27 A g⁻¹	90	1000	2 A g⁻¹	[21]
LiV$_3$O$_8$ network	Es	321	0.1 A g⁻¹	203	2 A g⁻¹	85	100	0.1 A g⁻¹	[22]
V$_3$O$_7$ NW/Gr scrolls	SSS	321	0.1 A g⁻¹	162	3 A g⁻¹	87	400	2 A g⁻¹	[23]
P-V$_2$O$_5$ NB ARs	Hy/An	142	0.05 A g⁻¹	130	1 A g⁻¹	99	100	1 A g⁻¹	[24]
P-Fe$_3$O$_4$/VO$_x$/Gr NWs	PSP	1164	0.1 A g⁻¹	500	5 A g⁻¹	99	100	2 A g⁻¹	[25]
V$_2$O$_5$/PEDOT/MnO$_2$	Hy/An	196	0.05 A g⁻¹	48	0.5 A g⁻¹	99	200	0.1 A g⁻¹	[26]
V$_6$O$_{13}$ nanotextiles	SA	326	0.02 A g⁻¹	134	0.5 A g⁻¹	80	100	0.5 A g⁻¹	[27]
LiV$_3$O$_8$ NWs	Hy/An	176	1.5 A g⁻¹	137	2 A g⁻¹	98	400	0.2 A g⁻¹	[28]
Mn$_3$O$_4$/VGCF NR	Stirring/calcine	950	0.2 A g⁻¹	390	5 A g⁻¹	–	–	–	[29]
GrNSs/MnO NWs	Es	995	0.05 A g⁻¹	541	0.5 A g⁻¹	–	–	–	[30]
MnO/C c-s NWs	Hy/An	1119	0.5 A g⁻¹	463	5 A g⁻¹	99	1000	2 A g⁻¹	[31]
CNF/MnO NC	Hy/An	936	0.2 A g⁻¹	472	0.8 A g⁻¹	98	150	0.2 A g⁻¹	[32]

(continued)

Table 3.1 (continued)

Materials	Synthesis method	Reversible Cp		Rate capacity		Capacity retention			References
		Cp	Cd	Cp	Cd	Pr%	Cycle	Cd	
CNF/CoMn$_2$O$_4$ NC	Hy/An	870	0.2 A g^{-1}	610	1.2 A g^{-1}	–	–	–	[32]
Mn$_2$O$_3$ NW	Hy/An	816	0.1 A g^{-1}	–		62	100	0.1 A g^{-1}	[33]
MnO$_x$/C y-s NRs	Sol–gel	660	0.1 A g^{-1}	634	0.5 A g^{-1}	73	900	0.5 A g^{-1}	[34]
MnO/ZnMn$_2$O$_4$ NRs	Carbonization	803	0.05 A g^{-1}	595	1 A g^{-1}	–	–	–	[35]
Ag NWs/γ–Fe$_2$O$_3$ NC	Oxidation/CA	890	0.1 A g^{-1}	550	2.0 A g^{-1}	–	–	–	[36]
Fe$_{0.12}$V$_2$O$_5$ NW ARs	Hy	382	0.03 A g^{-1}	–		86	500	0.5 A g^{-1}	[37]
N–C Fe$_3$O$_4$/SnO$_2$ NFs	Es	1223	0.1 A g^{-1}	640	0.8 A g^{-1}	70	80	0.1 A g^{-1}	[38]
LBIOX NT	Bacterial T	900	0.05 C	550	1 C	–	–	–	[39]
Fe$_2$O$_3$/C NFs	Es	913	0.5 A g^{-1}	913	5.0 A g^{-1}	84	300	1 A g^{-1}	[40]
NiFe$_2$O$_4$/Fe$_2$O$_3$ NT	Annealing	1393	0.1 A g^{-1}	424	2 A g^{-1}	67	100	0.1 A g^{-1}	[41]
α-Fe$_2$O$_3$ NR ARs	SPG	801	5 A g^{-1}	499	20 A g^{-1}	–	–	–	[42]
Fe$_3$O$_4$–Fe NWs	DIC	1012	1 C	500	20 C	94	100	1 C	[43]
Fe$_2$O$_3$ NRs/CNT-Gf	Hy	1000	0.2 A g^{-1}	500	3 A g^{-1}	–	–	–	[44]
Fe$_3$O4/VOx/Gr NWs	PS	1146	0.1 A g^{-1}	500	5 A g^{-1}	–	–	–	[25]
CoMoO$_4$/PPy c-s NW	Hy/CPP	1425	0.1 A g^{-1}	753	1.2 A g^{-1}	89	100	0.1 A g^{-1}	[45]
P-FeCo$_2$O$_4$ NNs AR	Hy	1962	0.1 A g^{-1}	875	2 A g^{-1}	68	200	0.1 A g^{-1}	[46]
P-NiCo$_2$O$_4$ NBs	Hy	1056	0.5 A g^{-1}	981	0.5 A g^{-1}	93	100	0.5 A g^{-1}	[47]
Zn/Co oxide ARs	Hy	804	0.5 Ag^{-1}	738	1 A g^{-1}	70	100	1 A g^{-1}	[48]
P-Co$_3$O$_4$ rods	BTS	936	0.24 A g^{-1}	–	–	97	10	0.2 A g^{-1}	[49]

(continued)

Table 3.1 (continued)

Materials	Synthesis method	Reversible Cp		Rate capacity		Capacity retention			References
		Cp	Cd	Cp	Cd	Pr%	Cycle	Cd	
CoO NWs Clusters	Hy/An	1516	1 C	1331	5 C	–	–	–	[50]
Co_3O_4 NTs	Es	856	0.25 C	677	1 C	90	60	1 C	[51]
$MnCo_2O_4$ NW ARs	Hy	966	0.1 A g^{-1}	345	1 A g^{-1}	93	50	0.1 A g^{-1}	[52]
$NiCoO_2$ NSs/CNT	Calcined	1309	0.4 A g^{-1}	933	0.8 A g^{-1}	–	–	–	[53]
NiO–CoO/C brushes	Hy	962	0.2 A g^{-1}	–		83	200	0.2 A g^{-1}	[54]
TiC/NiO c-s NWs	Hy	508	0.2 A g^{-1}	368	3 A g^{-1}	90	60	0.2 A g^{-1}	[55]
Mp $NiCo_2O_4$ NW ARs	Hy/An	1012	0.5 A g^{-1}	778	2 A g^{-1}	84	100	0.5 A g^{-1}	[56]
MWCNT/NiO ARs	CVD	958	0.1 C	820	1 C	–	–	–	[57]
NiO/CNT tubes	Solution route	1013	0.1 A g^{-1}	775	0.8 A g^{-1}	–	–	–	[58]
UGF/CNTs/NiO NTs	SS	1097	0.2 A g^{-1}	717	2 A g^{-1}	–	–	–	[59]
$NiCo_2O_4$ NW/CFC	Hy/An	1092	0.5 A g^{-1}	507	4 A g^{-1}	92	100	0.5 A g^{-1}	[60]
$Ni_xCo_{3-x}O_4$ H-MT	Hy/OR	1470	0.8 A g^{-1}	1100	4 A g^{-1}	42	1000	0.8 A g^{-1}	[61]
C–$NiCo_2O_4/SnO_2$ NB	Hy	1150	0.1 A g^{-1}	600	0.6 A g^{-1}	57	100	0.1 A g^{-1}	[62]
Ni–Co oxide prisms	Solvothermal	1025	0.2 A g^{-1}	535	0.8 A g^{-1}	96	30	0.2 A g^{-1}	[63]
CuO/C c-s NW ARs	ED/ALD	650	0.5 C	425	3 C	85	290	3 C	[64]
Leaf-like Mp CuO	Decomposition	753	0.1 A g^{-1}	495	2 A g^{-1}	99	300	0.5 A g^{-1}	[65]
CuO NWs/GrQDs	SM/An	780	0.33 C	330	30 C	80	300	0.33 C	[66]
Fe_3O_4/CuO NWs	Anodization	734	0.41 A g^{-1}	319	8.2 A g^{-1}	–	–	–	[67]
$CuGeO_3$ NWs/RGO	Hy	1000	0.1 A g^{-1}	879	1 A g^{-1}	78	120	0.1 A g^{-1}	[68]

(continued)

Table 3.1 (continued)

Materials	Synthesis method	Reversible Cp		Rate capacity		Capacity retention			References
		Cp	Cd	Cp	Cd	Pr%	Cycle	Cd	
CuO NWs on Ni foam	TO	692	$0.1\ A\ g^{-1}$	449	$1\ A\ g^{-1}$	96	600	$1\ A\ g^{-1}$	[69]
CuO/LiNi$_{0.5}$Mn$_{1.5}$O$_4$	LSR	660	0.1 C	240	10 C	84	100	0.5 C	[70]
CuO/CoO c-s ARs	Heat treatment	1364	$0.1\ A\ g^{-1}$	1140	$1.0\ A\ g^{-1}$	–			[71]
CuO NPs in Cu NTs	TO	600	$0.1\ A\ g^{-1}$	175	$15\ A\ g^{-1}$	94	200	$15\ A\ g^{-1}$	[72]
P–CuO NTs/Gr	MAP	725	$0.1\ A\ g^{-1}$	501	$0.5\ A\ g^{-1}$	81	250	$0.5\ A\ g^{-1}$	[73]
3D CuO/CNx c-s NC	RFMS	804	$0.05\ A\ g^{-1}$	554	$2.5\ A\ g^{-1}$	83	200	$0.05\ A\ g^{-1}$	[74]
Zn$_2$GeO$_4$/C NWs	CVD	1162	$0.2\ A\ g^{-1}$	465	$10\ A\ g^{-1}$	–	–	–	[75]
SnO$_2$ NRs/ZnO NFs	Es/calcination	748	$0.08\ A\ g^{-1}$	314	$0.8\ A\ g^{-1}$	–	–	–	[76]
Zn$_2$SnO$_4$ NWs	VTM	1000	$0.12\ A\ g^{-1}$	–		69	60	$0.1\ A\ g^{-1}$	[77]
ZnO/ZnO QDs/C NR	IEP	1055	$0.1\ A\ g^{-1}$	530	$1\ A\ g^{-1}$	89	100	$0.5\ A\ g^{-1}$	[78]
ZnO/Cu on CNFs	Es	812	$0.1\ A\ g^{-1}$	493	$1\ A\ g^{-1}$	–	–	–	[79]
ZnO–C–rGO NFs	Es	815	$0.05\ A\ g^{-1}$	472	$0.5\ A\ g^{-1}$	80	100	$0.05\ A\ g^{-1}$	[80]
NiO/ZnO NFs	Es	949	$0.2\ A\ g^{-1}$	707	$3.2\ A\ g^{-1}$	–	–	–	[81]
CNT/MoO$_{3-x}$ NB	Hy	421	$0.2\ A\ g^{-1}$	293	$2\ A\ g^{-1}$	–	–	–	[82]
α-MoO$_3$ NB	Hy	1000	$0.05\ A\ g^{-1}$	443	$2\ A\ g^{-1}$	–	–	–	[83]

[107]. At present, the breakthroughs made in these field have been chiefly based on a physical capture method for the polysulfide and sulfur species. Simultaneously, attention has also been paid to utilizing chemical routes to restrain the polysulfide shuttle. Furthermore, some transition metal oxides (e.g., TiO_2, La_2O_3, Al_2O_3, $Mg_{0.6}Ni_{0.4}O$) as functional absorbents have been shown to be efficient at adsorbing Li polysulfides, thus enhancing the long-term cyclability of Li–S batteries [108–110].

3.3   Sodium-Ion Batteries

Sodium-ion batteries are a kind of rechargeable metal-ion battery that employs Na^+ as a charge carrier. Sodium-ion batteries are promising devices with the potential to meet the demands of large-scale energy storage, such as grid-level storage, because sodium is inexpensive and earth abundant [111]. In addition, sodium-ion batteries have an analogous rocking chair mechanism to that of lithium-ion batteries, offering a long cycling life and high reversibility. Furthermore, Na has a suitable redox potential $\left(E^o_{(Na+/Na)} = -2.71 \text{ V}\right)$, which is merely 0.3 V above that of Li, meaning there is a lower energy cost because rechargeable batteries based on Na possess more promise for electrochemical energy storage applications [112].

3.4   Metal-Air Batteries

The lithium-air battery is a type of metal-air battery or electrochemical cell that utilizes the reduction of O_2 at the cathode and oxidation of Li at the anode to induce an electric current [113]. Zinc-air batteries are a century-old technology but have attracted increasing interest recently [114]. The theoretical energy density of zinc-air batteries is 1086 Wh kg^{-1}, which is approximately 5 times higher than that of lithium-ion batteries at present. Furthermore, zinc-air batteries could be fabricated at a low cost of under $10 kW^{-1} h^{-1}, which is approximately two orders of magnitude less than lithium batteries [8]. For many devices, zinc-air batteries may provide the highest energy density of any main battery system. Some zinc-air batteries have been successfully applied to the telecommunications and medical fields. Although they have high energy densities, they have a low power output of under 10 mW due to the poor efficiency of air catalysts. Moreover, the development of a rechargeable zinc-air battery system with an enhanced cycling ability has been affected by non-uniform zinc deposition and dissolution and the lack of excellent bifunctional catalysts. This process general proceeds until they become oversaturated with the electrolyte solution, after which the zincate ions transform to undissolved zinc oxide, as shown below:

Negative electrode $\quad 2Zn + 8OH^- = 2Zn(OH)_4^{2-} + 4e^-$
$\qquad\qquad\qquad\qquad 2Zn(OH)_4^{2-} = 2ZnO + 2H_2O + 4OH^-$
Positive electrode $\quad O_2 + 4e^- + 2H_2O = 4OH^-$
Overall reaction $\qquad 2Zn + O_2 = 2ZnO$
Parasitic reaction $\qquad Zn + 2H_2O = Zn(OH)_2 + H_2$

A possible alternative to the lithium-air battery is the sodium-air battery because of its abundant natural resources and the potential of 2.33 (V) for sodium. Compared with lithium-ion batteries, sodium-air batteries could provide a higher theoretical specific charge of 1605 Wh kg^{-1}, which is an order of magnitude higher than the theoretical value of olivine electrode materials, such as $LiFePO_4$ (approx. 170 Ah kg^{-1}). One challenge of sodium-air batteries that must be address is how to enable the typical reversible reactions (i.e., $4Na + O_2 \leftrightharpoons 2Na_2O$, $2Na + O_2 \leftrightharpoons Na_2O_2$, and $Na + O_2 \leftrightharpoons NaO_2$) relative to aprotic electrolytes. Therefore, different catalysts and additives must be explored to find high-efficiency cathode materials of sodium-air batteries.

3.5 Other Advanced Batteries

Redox-flow batteries are originated in the 1960s, with the use of the chlorine/zinc hydrate battery. A redox-flow battery utilizes two circulating soluble redox couples as electroactive materials that are gradually oxidized to decrease energy delivery or storage. A variety of redox-flow battery devices has been researched since the 1970s. A partial list contains all-vanadium, iron/chromium, zinc-cerium vanadium/bromine, zinc/bromine and bromine/polysulfide. In comparison, the zinc/bromine (1.85 V) and all-vanadium (1.26 V) devices are the most promising and have demonstrated energy storage. Redox-flow battery devices have numerous advantages (47). The simplicity of the reactions on the electrode is comparable to various traditional batteries and involves the electrode morphology, changes in the electrolyte degradation and phase transformations. One of the most attractive features of redox-flow batteries is that the energy and power are uncoupled, a characteristic that is not available in numerous other energy storage devices. This also gives higher design elasticity for established energy storage equipment. The speculated capacity can be enhanced by easily increasing the concentration of the electrolyte or increasing the magnitude of the liquid accumulator holding the reactants. Most importantly, the power of the redox-flow battery system could be tuned by the following points:

(1) Connecting stacks in either series or parallel configurations;
(2) Using bipolar electrodes;
(3) Modifying the amounts of cells in the stacks.

Potassium-ion batteries are novel energy storing devices that have various advantages such as similar output voltages to those of Li-ion batteries and abundant precursor materials (the elemental reserves of K are close to those of Na) [115–117].

The low cost of K is a key advantage, showing potassium batteries as promising candidates for large-scale energy storage devices such as those needed for electric vehicles and household energy storage applications [5]. Furthermore, potassium-ion batteries show greater potential to achieve a faster charging rate compared to lithium-ion batteries, meaning that next-generation mobile phones based on potassium-ion batteries may be charged in several minute. The diffusion coefficient of potassium ions in potassium-ion batteries is higher than that of lithium ions in lithium-ion batteries because of the smaller Stokes radius of solvated potassium ions. Because the electrochemical potential of potassium ions is similar to that of lithium ions, the cell potential is parallel to that of lithium-ion cell. Potassium-ion batteries could show a greater range of cathode active materials, which could provide a rechargeability with a lower cost. Nevertheless, due to the large radius of potassium ions, most of cathodes are unable to adapt to the architectural deterioration during insertion and extraction processes, generating a low capacity and short-term cycling stability.

Magnesium batteries have been regarded as a promising technology because of magnesium's high volumetric capacity, safety and abundance. Nevertheless, very few materials show reversible performance for magnesium-ion systems. Magnesium, compared to lithium, is an anode material has a theoretical energy density per unit mass of $18.8 \, MJ \, kg^{-1}$, which is under half that of lithium ($42.3 \, MJ \, kg^{-1}$). Nevertheless, magnesium possesses a volumetric energy density of $32.731 \, GJ \, m^{-3}$, which is approximately 50% higher than the volumetric energy density of lithium (approximately $22.569 \, GJ \, m^{-3}$). Furthermore, magnesium electrodes have not shown the formation of dendritic crystals during the charging process, which could permit magnesium to be used without the intercalation of a chemical compound at the anode. The advantage of not needing an intercalation layer for the magnesium anode may increase the supreme relative volume energy density to approximately five times that of a lithium-ion battery.

Battery-supercapacitor hybrid devices are constructed with a high-rate capacitive-type electrode and a high-capacity battery electrode, and they have attracted wide attention because of their promising applications in miniaturized optoelectronic/electronic equipment, smart electric grids, electric vehicles, etc. [11]. With a reasonable design, battery-supercapacitor hybrid devices will offer fascinating advantages including environmental friendliness, low cost, high performance and safety.

References

1. Lu J, Chen Z, Ma Z, Pan F, Curtiss LA, Amine K (2016) The role of nanotechnology in the development of battery materials for electric vehicles. Nat Nanotechnol 11(12):1031–1038. https://doi.org/10.1038/nnano.2016.207
2. Ellingsen LA, Hung CR, Majeau-Bettez G, Singh B, Chen Z, Whittingham MS, Stromman AH (2016) Nanotechnology for environmentally sustainable electromobility. Nat Nanotechnol 11(12):1039–1051. https://doi.org/10.1038/nnano.2016.237

3. Ni Q, Bai Y, Wu F, Wu C (2017) Polyanion-type electrode materials for sodium-ion batteries. Adv Sci 4(3):1600275. https://doi.org/10.1002/advs.201600275

4. Reddy MV, Subba Rao GV, Chowdari BV (2013) Metal oxides and oxysalts as anode materials for Li ion batteries. Chem Rev 113(7):5364–5457. https://doi.org/10.1021/cr3001884

5. Wang X, Xu X, Niu C, Meng J, Huang M, Liu X, Liu Z, Mai L (2017) Earth abundant Fe/Mn-based layered oxide interconnected nanowires for advanced K-ion full batteries. Nano Lett 17(1):544–550. https://doi.org/10.1021/acs.nanolett.6b04611

6. Liu X, Huang JQ, Zhang Q, Mai L (2017) Nanostructured metal oxides and sulfides for lithium–sulfur batteries. Adv Mater 29(20). https://doi.org/10.1002/adma.201601759

7. Li B, Gu M, Nie Z, Wei X, Wang C, Sprenkle V, Wang W (2014) Nanorod niobium oxide as powerful catalysts for an all vanadium redox flow battery. Nano Lett 14(1):158–165. https://doi.org/10.1021/nl403674a

8. Li Y, Dai H (2014) Recent advances in zinc-air batteries. Chem Soc Rev 43(15):5257–5275. https://doi.org/10.1039/c4cs00015c

9. Jian Z, Liu P, Li F, He P, Guo X, Chen M, Zhou H (2014) Core–shell-structured CNT@RuO_2 composite as a high-performance cathode catalyst for rechargeable Li–O_2 batteries. Angew Chem Int Edit 53(2):442–446. https://doi.org/10.1002/anie.201307976

10. Prabu M, Ramakrishnan P, Ganesan P, Manthiram A, Shanmugam S (2015) LaTi0.65Fe0.35O3−δ nanoparticle-decorated nitrogen-doped carbon nanorods as an advanced hierarchical air electrode for rechargeable metal-air batteries. Nano Energy 15:92–103. https://doi.org/10.1016/j.nanoen.2015.04.005

11. Zuo W, Li R, Zhou C, Li Y, Xia J, Liu J (2017) Battery-supercapacitor hybrid devices: recent progress and future prospects. Adv Sci 4(7):1600539. https://doi.org/10.1002/advs.201600539

12. Grey CP, Tarascon JM (2017) Sustainability and in situ monitoring in battery development. Nat Mater 16(1):45–56. https://doi.org/10.1038/nmat4777

13. Goriparti S, Miele E, Prato M, Scarpellini A, Marras S, Monaco S, Toma A, Messina GC, Alabastri A, De Angelis F, Manna L, Capiglia C, Zaccaria RP (2015) Direct synthesis of carbon-doped TiO_2-bronze nanowires as anode materials for high performance lithium-ion batteries. ACS Appl Mater Inter 7(45):25139–25146. https://doi.org/10.1021/acsami.5b06426

14. Zhang Y, Meng Y, Zhu K, Qiu H, Ju Y, Gao Y, Du F, Zou B, Chen G, Wei Y (2016) Copper–doped titanium dioxide bronze nanowires with superior high rate capability for lithium ion batteries. ACS Appl Mater Inter 8(12):7957–7965. https://doi.org/10.1021/acsami.5b10766

15. Chen J, Song W, Hou H, Zhang Y, Jing M, Jia X, Ji X (2015) Ti^{3+} self-doped dark rutile TiO_2 ultrafine nanorods with durable high-rate capability for lithium-ion batteries. Adv Funct Mater 25(43):6793–6801. https://doi.org/10.1002/adfm.201502978

16. Liu Y, Elzatahry AA, Luo W, Lan K, Zhang P, Fan J, Wei Y, Wang C, Deng Y, Zheng G, Zhang F, Tang Y, Mai L, Zhao D (2016) Surfactant-templating strategy for ultrathin mesoporous TiO_2 coating on flexible graphitized carbon supports for high-performance lithium-ion battery. Nano Energy 25:80–90. https://doi.org/10.1016/j.nanoen.2016.04.028

17. Zhu C, Xia X, Liu J, Fan Z, Chao D, Zhang H, Fan HJ (2014) TiO_2 nanotube@SnO_2 nanoflake core–branch arrays for lithium-ion battery anode. Nano Energy 4:105–112. https://doi.org/10.1016/j.nanoen.2013.12.018

18. Mao M, Mei L, Guo D, Wu L, Zhang D, Li Q, Wang T (2014) High electrochemical performance based on the TiO_2 nanobelt@few-layered MoS_2 structure for lithium-ion batteries. Nanoscale 6(21):12350–12353. https://doi.org/10.1039/c4nr03991b

19. Chao D, Xia X, Liu J, Fan Z, Ng CF, Lin J, Zhang H, Shen ZX, Fan HJ (2014) A V_2O_5/conductive-polymer core/shell nanobelt array on three-dimensional graphite foam: a high-rate, ultrastable, and freestanding cathode for lithium-ion batteries. Adv Mater 26(33):5794–5800. https://doi.org/10.1002/adma.201400719

20. Huang SZ, Cai Y, Jin J, Li Y, Zheng XF, Wang HE, Wu M, Chen LH, Su BL (2014) Annealed vanadium oxide nanowires and nanotubes as high performance cathode materials for lithium ion batteries. J Mater Chem A 2(34):14099. https://doi.org/10.1039/c4ta02339k

21. Ren G, Hoque MNF, Pan X, Warzywoda J, Fan Z (2015) Vertically aligned $VO_2(B)$ nanobelt forest and its three-dimensional structure on oriented graphene for energy storage. J Mater Chem A 3(20):10787–10794. https://doi.org/10.1039/c5ta01900a

22. Ren W, Zheng Z, Luo Y, Chen W, Niu C, Zhao K, Yan M, Zhang L, Meng J, Mai L (2015) An electrospun hierarchical LiV_3O_8 nanowire-in-network for high-rate and long-life lithium batteries. J Mater Chem A 3(39):19850–19856. https://doi.org/10.1039/c5ta04643b

23. Yan M, Wang F, Han C, Ma X, Xu X, An Q, Xu L, Niu C, Zhao Y, Tian X, Hu P, Wu H, Mai L (2013) Nanowire templated semihollow bicontinuous graphene scrolls: designed construction, mechanism, and enhanced energy storage performance. J Am Chem Soc 135(48):18176–18182. https://doi.org/10.1021/ja409027s

24. Qin M, Liang Q, Pan A, Liang S, Zhang Q, Tang Y, Tan X (2014) Template-free synthesis of vanadium oxides nanobelt arrays as high-rate cathode materials for lithium ion batteries. J Power Sources 268:700–705. https://doi.org/10.1016/j.jpowsour.2014.06.103

25. An Q, Lv F, Liu Q, Han C, Zhao K, Sheng J, Wei Q, Yan M, Mai L (2014) Amorphous vanadium oxide matrixes supporting hierarchical porous Fe_3O_4/graphene nanowires as a high-rate lithium storage anode. Nano Lett 14(11):6250–6256. https://doi.org/10.1021/nl5025694

26. Mai L, Dong F, Xu X, Luo Y, An Q, Zhao Y, Pan J, Yang J (2013) Cucumber-like V_2O_5/poly(3,4-ethylenedioxythiophene)&MnO_2 nanowires with enhanced electrochemical cyclability. Nano Lett 13(2):740–745. https://doi.org/10.1021/nl304434v

27. Ding YL, Wen Y, Wu C, van Aken PA, Maier J, Yu Y (2015) 3D V_6O_{13} nanotextiles assembled from interconnected nanogrooves as cathode materials for high-energy lithium ion batteries. Nano Lett 15(2):1388–1394. https://doi.org/10.1021/nl504705z

28. Xu X, Luo YZ, Mai LQ, Zhao YL, An QY, Xu L, Hu F, Zhang L, Zhang QJ (2012) Topotactically synthesized ultralong LiV_3O_8 nanowire cathode materials for high-rate and long-life rechargeable lithium batteries. NPG Asia Mater 4(6):e20. https://doi.org/10.1038/am.2012.36

29. Ma F, Yuan A, Xu J (2014) Nanoparticulate Mn_3O_4/VGCF composite conversion-anode material with extraordinarily high capacity and excellent rate capability for lithium ion batteries. ACS Appl Mater Inter 6(20):18129–18138. https://doi.org/10.1021/am505022u

30. Sun Q, Wang Z, Zhang Z, Yu Q, Qu Y, Zhang J, Yu Y, Xiang B (2016) Rational design of graphene-reinforced mno nanowires with enhanced electrochemical performance for Li-ion batteries. ACS Appl Mater Inter 8(10):6303–6308. https://doi.org/10.1021/acsami.6b00122

31. Jiang H, Hu Y, Guo S, Yan C, Lee PS, Li C (2014) Rational design of MnO/Carbon nanopeapods with internal void space for high-rate and long-life Li-ion batteries. ACS Nano 8(6):6038–6046

32. Zhang G, Wu HB, Hoster HE, Lou XW (2014) Strongly coupled carbon nanofiber–metal oxide coaxial nanocables with enhanced lithium storage properties. Energ Environ Sci 7(1):302–305. https://doi.org/10.1039/c3ee43123a

33. Wang Y, Wang Y, Jia D, Peng Z, Xia Y, Zheng G (2014) All-nanowire based Li-ion full cells using homologous Mn_2O_3 and $LiMn_2O_4$. Nano Lett 14(2):1080–1084. https://doi.org/10.1021/nl4047834

34. Cai Z, Xu L, Yan M, Han C, He L, Hercule KM, Niu C, Yuan Z, Xu W, Qu L, Zhao K, Mai L (2015) Manganese oxide/carbon yolk–shell nanorod anodes for high capacity lithium batteries. Nano Lett 15(1):738–744. https://doi.org/10.1021/nl504427d

35. Zhong M, Yang D, Xie C, Zhang Z, Zhou Z, Bu XH (2016) Yolk–shell $MnO@ZnMn_2O_4$/N–C nanorods derived from α-MnO_2/ZIF-8 as anode materials for lithium ion batteries. Small 12(40):5564–5571. https://doi.org/10.1002/smll.201601959

36. Geng H, Ge D, Lu S, Wang J, Ye Z, Yang Y, Zheng G, Gu H (2015) Preparation of a gamma-Fe_2O_3/Ag nanowire coaxial nanocable for high-performance lithium-ion batteries. Chem-Eur J 21(31):11129–11133. https://doi.org/10.1002/chem.201500819

37. Cao Y, Fang D, Liu R, Jiang M, Zhang H, Li G, Luo Z, Liu X, Xu J, Xu W, Xiong C (2015) Three-dimensional porous iron vanadate nanowire arrays as a high-performance lithium-ion battery. ACS Appl Mater Inter 7(50):27685–27693. https://doi.org/10.1021/acsami.5b08282

38. Xie W, Li S, Wang S, Xue S, Liu Z, Jiang X, He D (2014) N-doped amorphous carbon coated Fe_3O_4/SnO_2 coaxial nanofibers as a binder-free self-supported electrode for lithium ion batteries. ACS Appl Mater Inter 6(22):20334–20339. https://doi.org/10.1021/am505829v

39. Hashimoto H, Kobayashi G, Sakuma R, Fujii T, Hayashi N, Suzuki T, Kanno R, Takano M, Takada J (2014) Bacterial nanometric amorphous Fe-based oxide: a potential lithium-ion battery anode material. ACS Appl Mater Inter 6(8):5374–5378. https://doi.org/10.1021/am500905y

40. Cho JS, Hong YJ, Kang YC (2015) Design and synthesis of bubble–nanorod-structured Fe_2O_3–Carbon nanofibers as advanced anode material for Li-ion batteries. ACS Nano 9(4):4026–4035

41. Huang G, Zhang F, Zhang L, Du X, Wang J, Wang L (2014) Hierarchical $NiFe_2O_4/Fe_2O_3$ nanotubes derived from metal organic frameworks for superior lithium ion battery anodes. J Mater Chem A 2(21):8048–8053. https://doi.org/10.1039/c4ta00200h

42. Chen S, Xin Y, Zhou Y, Zhang F, Ma Y, Zhou H, Qi L (2015) Robust α-Fe_2O_3 nanorod arrays with optimized interstices as high-performance 3D anodes for high-rate lithium ion batteries. J Mater Chem A 3(25):13377–13383. https://doi.org/10.1039/c5ta02089a

43. Cheng K, Yang F, Ye K, Zhang Y, Jiang X, Yin J, Wang G, Cao D (2014) Highly porous Fe_3O_4–Fe nanowires grown on C/TiC nanofiber arrays as the high performance anode of lithium-ion batteries. J Power Sources 258:260–265. https://doi.org/10.1016/j.jpowsour.2014.02.038

44. Chen M, Liu J, Chao D, Wang J, Yin J, Lin J, Fan H, Shen Z (2014) Porous α-Fe_2O_3 nanorods supported on carbon nanotubes-graphene foam as superior anode for lithium ion batteries. Nano Energy 9:364–372. https://doi.org/10.1016/j.nanoen.2014.08.011

45. Chen Y, Liu B, Jiang W, Liu Q, Liu J, Wang J, Zhang H, Jing X (2015) Coaxial three-dimensional $CoMoO_4$ nanowire arrays with conductive coating on carbon cloth for high-performance lithium ion battery anode. J Power Sources 300:132–138. https://doi.org/10.1016/j.jpowsour.2015.09.011

46. Liu L, Zhang H, Mu Y, Yang J, Wang Y (2016) Porous iron cobaltate nanoneedles array on nickel foam as anode materials for lithium-ion batteries with enhanced electrochemical performance. ACS Appl Mater Inter 8(2):1351–1359. https://doi.org/10.1021/acsami.5b10237

47. Mondal AK, Su D, Chen S, Xie X, Wang G (2014) Highly porous $NiCo_2O_4$ Nanoflakes and nanobelts as anode materials for lithium-ion batteries with excellent rate capability. ACS Appl Mater Inter 6(17):14827–14835. https://doi.org/10.1021/am5036913

48. Yu J, Chen S, Hao W, Zhang S (2016) Fibrous-root-inspired design and lithium storage applications of a Co–Zn binary synergistic nanoarray system. ACS Nano 10(2):2500–2508. https://doi.org/10.1021/acsnano.5b07352

49. Shim HW, Jin YH, Seo SD, Lee SH, Kim DW (2011) Highly reversible lithium storage in bacillus subtilis-directed porous Co_3O_4 nanostructures. ACS Nano 5:443

50. Cao K, Jiao L, Liu Y, Liu H, Wang Y, Yuan H (2015) Ultra-high capacity lithium-ion batteries with hierarchical CoO nanowire clusters as binder free electrodes. Adv Funct Mater 25(7):1082–1089. https://doi.org/10.1002/adfm.201403111

51. Chen M, Xia X, Yin J, Chen Q (2015) Construction of Co_3O_4 nanotubes as high-performance anode material for lithium ion batteries. Electrochim Acta 160:15–21. https://doi.org/10.1016/j.electacta.2015.02.055

52. Li L, Zhang YQ, Liu XY, Shi SJ, Zhao XY, Zhang H, Ge X, Cai GF, Gu CD, Wang XL, Tu JP (2014) One-dimension $MnCo_2O_4$ nanowire arrays for electrochemical energy storage. Electrochim Acta 116:467–474. https://doi.org/10.1016/j.electacta.2013.11.081

53. Xu X, Dong B, Ding S, Xiao C, Yu D (2014) Hierarchical $NiCoO_2$ nanosheets supported on amorphous carbon nanotubes for high-capacity lithium-ion batteries with a long cycle life. J Mater Chem A 2(32):13069. https://doi.org/10.1039/c4ta02003k

54. Wei Y, Yan F, Tang X, Luo Y, Zhang M, Wei W, Chen L (2015) Solvent-controlled synthesis of NiO–CoO/carbon fiber nanobrushes with different densities and their excellent properties for lithium ion storage. ACS Appl Mater Inter 7(39):21703–21711. https://doi.org/10.1021/acsami.5b07233

55. Huang H, Feng T, Gan Y, Fang M, Xia Y, Liang C, Tao X, Zhang W (2015) TiC/NiO core/shell nanoarchitecture with battery-capacitive synchronous lithium storage for high-performance lithium-ion battery. ACS Appl Mater Inter 7(22):11842–11848. https://doi.org/10.1021/acsami.5b01372

56. Shen L, Che Q, Li H, Zhang X (2014) Mesoporous $NiCo_2O_4$ nanowire arrays grown on carbon textiles as binder-free flexible electrodes for energy storage. Adv Funct Mater 24(18):2630–2637. https://doi.org/10.1002/adfm.201303138

57. Susantyoko RA, Wang X, Xiao Q, Fitzgerald E, Zhang Q (2014) Sputtered nickel oxide on vertically-aligned multiwall carbon nanotube arrays for lithium-ion batteries. Carbon 68:619–627. https://doi.org/10.1016/j.carbon.2013.11.041

58. Xu X, Tan H, Xi K, Ding S, Yu D, Cheng S, Yang G, Peng X, Fakeeh A, Kumar RV (2015) Bamboo-like amorphous carbon nanotubes clad in ultrathin nickel oxide nanosheets for lithium-ion battery electrodes with long cycle life. Carbon 84:491–499. https://doi.org/10.1016/j.carbon.2014.12.040

59. Liu W, Lu C, Wang X, Liang K, Tay BK (2015) In situ fabrication of three-dimensional, ultra-thin graphite/carbon nanotube/NiO composite as binder-free electrode for high-performance energy storage. J Mater Chem A 3(2):624–633. https://doi.org/10.1039/c4ta04023f

60. Mo Y, Ru Q, Chen J, Song X, Guo L, Hu S, Peng S (2015) Three-dimensional $NiCo_2O_4$ nanowire arrays: preparation and storage behavior for flexible lithium-ion and sodium-ion batteries with improved electrochemical performance. J Mater Chem A 3(39):19765–19773. https://doi.org/10.1039/c5ta05931c

61. Zhou Y, Liu Y, Zhao W, Wang H, Li B, Zhou X, Shen H (2015) Controlled synthesis of series $NixCo_{3-x}O_4$ products: morphological evolution towards quasi-single-crystal structure for high-performance and stable lithium-ion batteries. Sci Rep 5:11584. https://doi.org/10.1038/srep11584

62. Gao G, Wu HB, Ding S, Lou XW (2015) Preparation of carbon-coated $NiCo_2O_4@SnO_2$ hetero-nanostructures and their reversible lithium storage properties. Small 11(4):432–436. https://doi.org/10.1002/smll.201400152

63. Yu L, Guan B, Xiao W, Lou XW (2015) Formation of yolk–shelled Ni–Co mixed oxide nanoprisms with enhanced electrochemical performance for hybrid supercapacitors and lithium ion batteries. Adv Energy Mater 5(21):1500981. https://doi.org/10.1002/aenm.201500981

64. Cao F, Xia XH, Pan GX, Chen J, Zhang YJ (2015) Construction of carbon nanoflakes shell on CuO nanowires core as enhanced core/shell arrays anode of lithium ion batteries. Electrochim Acta 178:574–579. https://doi.org/10.1016/j.electacta.2015.08.055

65. Bai Z, Zhang Y, Zhang Y, Guo C, Tang B (2015) A large-scale, green route to synthesize of leaf-like mesoporous CuO as high-performance anode materials for lithium ion batteries. Electrochim Acta 159(5):29–34

66. Zhu C, Chao D, Sun J, Bacho IM, Fan Z, Ng CF, Xia X, Huang H, Zhang H, Shen ZX (2015) Enhanced lithium storage performance of CuO nanowires by coating of graphene quantum dots. Adv Funct Mater 2(2):239–245

67. Saadat S, Zhu J, Sim D, Hng H, Yazami R, Yan Q (2013) Coaxial Fe_3O_4/CuO hybrid nanowires as ultra fast charge/discharge lithium-ion battery anodes. J Mater Chem A 1(30):8672–8678

68. Chen Z, Yan Y, Xin S, Li W, Qu J, Guo YG, Song WG (2013) Copper germanate nanowire/reduced graphene oxide anode materials for high energy lithium-ion batteries. J Mater Chem A 1(37):11404–11409

69. Zhang Q, Wang J, Xu D, Wang Z, Li X, Zhang K (2014) Facile large-scale synthesis of vertically aligned CuO nanowires on nickel foam: growth mechanism and remarkable electrochemical performance. J Mater Chem A 2(11):3865–3874

70. Zhang W, Ma G, Gu H, Yang Z, Cheng H (2015) A new lithium-ion battery: CuO nanorod array anode versus spinel $LiNi_{0.5}Mn_{1.5}O_4$ cathode. J Power Sources 273:561–565. https://doi.org/10.1016/j.jpowsour.2014.09.135

71. Wang J, Zhang Q, Li X, Zhang B, Mai L, Zhang K (2015) Smart construction of three-dimensional hierarchical tubular transition metal oxide core/shell heterostructures with high-capacity and long-cycle-life lithium storage. Nano Energy 12:437–446. https://doi.org/10.1016/j.nanoen.2015.01.003

72. Zhao Y, Mu S, Sun W, Liu Q, Li Y, Yan Z, Huo Z, Liang W (2016) Growth of copper oxide nanocrystals in metallic nanotubes for high performance battery anodes. Nanoscale 8(48):19994

73. Xiao S, Pan D, Wang L, Zhang Z, Lyu Z, Dong W, Chen X, Zhang D, Chen W, Li H (2016) Porous CuO nanotubes/graphene with sandwich architecture as high-performance anodes for lithium-ion batteries. Nanoscale 8(46):19343–19351

74. Tan G, Wu F, Yuan Y, Chen R, Zhao T, Yao Y, Qian J, Liu J, Ye Y, Shahbazian-Yassar R (2016) Freestanding three-dimensional core–shell nanoarrays for lithium-ion battery anodes. Nat Commun 7:11774

75. Chen W, Lu L, Maloney S, Yang Y, Wang W (2015) Coaxial Zn_2GeO_4@carbon nanowires directly grown on Cu foils as high-performance anodes for lithium ion batteries. Phys Chem Chem Phys 17(7):5109–5114. https://doi.org/10.1039/c4cp05705h

76. Zhu J, Zhang G, Gu S, Lu B (2014) SnO_2 nanorods on ZnO nanofibers: a new class of hierarchical nanostructures enabled by electrospinning as anode material for high-performance lithium-ion batteries. Electrochim Acta 150:308–313. https://doi.org/10.1016/j.electacta.2014.10.149

77. Cherian CT, Zheng M, Reddy MV, Chowdari BV, Sow CH (2013) Zn_2SnO_4 nanowires versus nanoplates: electrochemical performance and morphological evolution during Li-cycling. ACS Appl Mater Inter 5(13):6054–6060. https://doi.org/10.1021/am400802j

78. Zhang G, Hou S, Zhang H, Zeng W, Yan F, Li CC, Duan H (2015) High-performance and ultra-stable lithium-ion batteries based on MOF-derived ZnO@ZnO quantum dots/C core–shell nanorod arrays on a carbon cloth anode. Adv Mater 27(14):2400–2405. https://doi.org/10.1002/adma.201405222

79. Shen X, Mu D, Chen S, Huang R, Wu F (2014) Electrospun composite of ZnO/Cu nanocrystals-implanted carbon fibers as an anode material with high rate capability for lithium ion batteries. J Mater Chem A 2(12):4309. https://doi.org/10.1039/c3ta14685e

80. Shilpa S, Basavaraja BM, Majumder SB, Sharma A (2015) Electrospun hollow glassy carbon–reduced graphene oxide nanofibers with encapsulated ZnO nanoparticles: a free standing anode for Li-ion batteries. J Mater Chem A 3(10):5344–5351. https://doi.org/10.1039/c4ta07220k

81. Qiao L, Wang X, Qiao L, Sun X, Li X, Zheng Y, He D (2013) Single electrospun porous NiO–ZnO hybrid nanofibers as anode materials for advanced lithium-ion batteries. Nanoscale 5(7):3037–3042. https://doi.org/10.1039/c3nr34103h

82. Ni J, Wang G, Yang J, Gao D, Chen J, Gao L, Li Y (2014) Carbon nanotube-wired and oxygen-deficient MoO_3 nanobelts with enhanced lithium-storage capability. J Power Sources 247:90–94. https://doi.org/10.1016/j.jpowsour.2013.08.068

83. Sun Y, Wang J, Zhao B, Cai R, Ran R, Shao Z (2013) Binder-free α-MoO_3 nanobelt electrode for lithium-ion batteries utilizing van der Waals forces for film formation and connection with current collector. J Mater Chem A 1(15):4736. https://doi.org/10.1039/c3ta01285a

84. Dunn B, Kamath H, Tarascon JM (2011) Electrical energy storage for the grid: a battery of choices. Science 334(6058):928–935. https://doi.org/10.1126/science.1212741

85. Liu S, Dong Y, Zhao C, Zhao Z, Yu C, Wang Z, Qiu J (2015) Nitrogen-rich carbon coupled multifunctional metal oxide/graphene nanohybrids for long-life lithium storage and efficient oxygen reduction. Nano Energy 12:578–587

86. Liu S, Wang Z, Yu C, Wu HB, Wang G, Dong Q, Qiu J, Eychmüller A, David Lou XW (2013) A flexible TiO_2(B)-based battery electrode with superior power rate and ultralong cycle life. Adv Mater 25(25):3462–3467

87. Palacin MR, de Guibert A (2016) Why do batteries fail? Science 351(6273):1253292. https://doi.org/10.1126/science.1253292

88. Goodenough JB, Park KS (2013) The Li-ion rechargeable battery: a perspective. J Am Chem Soc 135(4):1167

89. Luo Y, Tang Y, Zheng S, Yan Y, Xue H, Pang H (2018) Dual anode materials for lithium- and sodium-ion batteries. J Mater Chem A

90. Choi JW, Aurbach D (2016) Promise and reality of post-lithium-ion batteries with high energy densities. Nat Rev Mater 1(4):16013. https://doi.org/10.1038/natrevmats.2016.13

91. Liu N, Lu Z, Zhao J, McDowell MT, Lee HW, Zhao W, Cui Y (2014) A pomegranate-inspired nanoscale design for large-volume-change lithium battery anodes. Nat Nanotechnol 9(3):187–192. https://doi.org/10.1038/nnano.2014.6

92. Sun Y, Liu N, Cui Y (2016) Promises and challenges of nanomaterials for lithium-based rechargeable batteries. Nat Energy 1(7):16071. https://doi.org/10.1038/nenergy.2016.71
93. El-Kady MF, Shao Y, Kaner RB (2016) Graphene for batteries, supercapacitors and beyond. Nat Rev Mater 1(7):16033. https://doi.org/10.1038/natrevmats.2016.33
94. Mei J, Liao T, Kou L, Sun Z (2017) Two-dimensional metal oxide nanomaterials for next-generation rechargeable batteries. Adv Mater. https://doi.org/10.1002/adma.201700176
95. Wang Z, Zhou L, Lou XW (2012) Metal oxide hollow nanostructures for lithium-ion batteries. Adv Mater 24(14):1903
96. Li D, Wang H, Liu HK, Guo Z (2016) A new strategy for achieving a high performance anode for lithium ion batteries-encapsulating germanium nanoparticles in carbon nanoboxes. Adv Energy Mater 6(5):1501666
97. Liu Z, Yu X, Paik U (2016) Etching-in-a-Box: a novel strategy to synthesize unique yolk–shelled Fe_3O_4@carbon with an ultralong cycling life for lithium storage. Adv Energy Mater 6(6):1502318
98. Pang Q, Liang X, Kwok CY, Nazar LF (2016) Advances in lithium–sulfur batteries based on multifunctional cathodes and electrolytes. Nat Energy 1(9):16132. https://doi.org/10.1038/nenergy.2016.132
99. Zhou G, Pei S, Li L, Wang DW, Wang S, Huang K, Yin LC, Li F, Cheng HM (2014) A graphene–pure-sulfur sandwich structure for ultrafast. Long-life lithium-sulfur batteries. Adv Mater 26(4):625–631
100. Evers S, Nazar LF (2013) New approaches for high energy density lithium–sulfur battery cathodes. Accounts Chem Res 46(5):1135
101. Manthiram A, Fu Y, Chung SH, Zu C, Su YS (2014) Rechargeable lithium–sulfur batteries. Chem Rev 114(23):11751–11787. https://doi.org/10.1021/cr500062v
102. Zhou G, Zhao Y, Zu C, Manthiram A (2015) Free-standing TiO_2 nanowire-embedded graphene hybrid membrane for advanced Li/dissolved polysulfide batteries. Nano Energy 12:240–249. https://doi.org/10.1016/j.nanoen.2014.12.029
103. Lv W, Li Z, Zhou G, Shao J, Kong D, Zheng X, Li B, Li F, Kang F, Yang Q (2014) Tailoring microstructure of graphene-based membrane by controlled removal of trapped water inspired by the phase diagram. Adv Funct Mater 24(22):3456–3463
104. Zhao MQ, Zhang Q, Huang JQ, Tian GL, Nie JQ, Peng HJ, Wei F (2014) Unstacked double-layer templated graphene for high-rate lithium–sulphur batteries. Nat Commun 5(5):3410
105. Zhang C, Lv W, Zhang W, Zheng X, Wu MB, Wei W, Tao Y, Li Z, Yang QH (2014) Reduction of graphene oxide by hydrogen sulfide: a promising strategy for pollutant control and as an electrode for Li–S batteries. Adv Energy Mater 4(7):1301565
106. Li Z, Jiang Y, Yuan L, Yi Z, Wu C, Liu Y, Strasser P, Huang Y (2014) A highly ordered meso@microporous carbon-supported sulfur@smaller sulfur core–shell structured cathode for Li–S batteries. ACS Nano 8(9):9295–9303
107. Cheng XB, Huang JQ, Zhang Q, Peng HJ, Zhao MQ, Wei F (2014) Aligned carbon nanotube/sulfur composite cathodes with high sulfur content for lithium–sulfur batteries. Nano Energy 4(2):65–72. https://doi.org/10.1016/j.nanoen.2013.12.013
108. Lin Z, Liang C (2015) Lithium–sulfur batteries: from liquid to solid cells. J Mater Chem A 3(3):936–958. https://doi.org/10.1039/c4ta04727c
109. Yang Y, Zheng G, Cui Y (2013) Nanostructured sulfur cathodes. Chem Soc Rev 42(7):3018–3032
110. Liang X, Kwok CY, Lodi-Marzano F, Pang Q, Cuisinier M, Huang H, Hart CJ, Houtarde D, Kaup K, Sommer H, Brezesinski T, Janek J, Nazar LF (2016) Tuning transition metal oxide–sulfur interactions for long life lithium sulfur batteries: the "Goldilocks" principle. Adv Energy Mater 6(6):1501636. https://doi.org/10.1002/aenm.201501636
111. Luo W, Shen F, Bommier C, Zhu H, Ji X, Hu L (2016) Na-ion battery anodes: materials and electrochemistry. Accounts Chem Res 49(2):231–240. https://doi.org/10.1021/acs.accounts.5b00482
112. Pan H, Hu Y-S, Chen L (2013) Room-temperature stationary sodium-ion batteries for large-scale electric energy storage. Energ Environ Sci 6(8):2338. https://doi.org/10.1039/c3ee40847g

113. Bruce PG, Freunberger SA, Hardwick LJ, Tarascon JM (2011) Li–O_2 and Li–S batteries with high energy storage. Nat Mater 11(1):19–29. https://doi.org/10.1038/nmat3191
114. Gu P, Zheng M, Zhao Q, Xiao X, Xue H, Pang H (2017) Rechargeable zinc–air batteries: a promising way to green energy. J Mater Chem A 5:7651–7666
115. Kundu D, Talaie E, Duffort V, Nazar LF (2015) The emerging chemistry of sodium ion batteries for electrochemical energy storage. Angew Chem Int Edit 54(11):3431–3448. https://doi.org/10.1002/anie.201410376
116. Jian Z, Luo W, Ji X (2015) Carbon electrodes for K-ion batteries. J Am Chem Soc 137(36):11566–11569
117. Share K, Cohn AP, Carter R, Rogers B, Pint CL (2016) Role of nitrogen-doped graphene for improved high-capacity potassium ion battery anodes. ACS Nano 10(10):9738

Chapter 4
One-Dimensional/One-Dimensional Analogue TMOs for Advanced Batteries

Abstract Transition metal oxides (TMOs) have a characteristic nature, including short-distance interactions between ions and carriers, rapid ionic-transport channels, and unique combinations of redox chemistry. Strikingly, due to these properties, TMOs have had great effects on the successful implementation of miscellaneous batteries. This chapter provide a comprehensive review of a variety of 1D/1D analogue–TMOs in batteries, including, titanium oxides, vanadium oxides, manganese oxides, iron oxides, cobalt oxides, nickel oxides, etc.

Keywords Transition metal oxides · One-dimensional architectures · One-dimensional analogue constructions · Batteries applications

The applications include lithium/sodium/potassium/magnesium-ion batteries, Li–S batteries, metal-air batteries, etc. [1]. The radar plots of the performance properties of different transition metal oxide nanomaterials is shown in Fig. 4.1.

4.1 Titanium Oxides

Titanium dioxides has many advantages, including cycling and structural stability; nevertheless, it has a poor theoretical capacity of only 178 mAh g^{-1}. Rutile TiO_2 nanorods, with a dark color from Ti^{3+}-doping and a mean diameter of approximately 7 nm, were first manufactured and applied as anodes of lithium-ion batteries [2]. HRTEM images of D–TiO_2 nanorods could be seen in Fig. 4.2a. This ultrafine nanorod shows a reversible specific capacitance of 185.7 mAh g^{-1} at 0.2 C and a remarkable cycling stability (98.4% retention after 1000 cycles at 50 C, as shown in Fig. 4.2b). The coulombic efficiency of Ti^{3+}-doped TiO_2 is approximately 10% greater than that of rutile TiO_2, which could be attributed to its speedy electron transfer due to the introduction of Ti^{3+}. The synergistic advantages are noticed when a shortened Li$^+$ diffusion distance is combined with a promoted electroconductivity resulting from the extra-fine nanorod construction, generating an extraordinary cycling stability and remarkable rate capability for applications in durable and fast

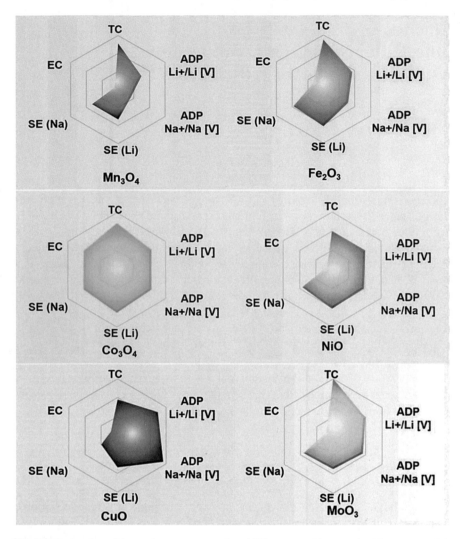

Fig. 4.1 Radar plots of the performance properties of different transition metal oxide nanomaterials. TC = theoretical capacity, ADP Li$^+$/Li [V] = average discharge potential Li$^+$/Li [V], ADP Na$^+$/Na [V] = average discharge potential Na$^+$/Na [V], SE (Li) = size expansion(Li), SE (Na) = size expansion (Na), EC = electrical conductivity

charge-discharge batteries. SnO_2 has a high theoretical capacity of 782 mAh g^{-1}, which is approximately twice that of the theoretical capacity of commercial graphite [3–6]. Fan et al. successfully designed SnO_2 nanoflakes on TiO_2 nanotube materials (Supplementary Fig. 4.2c) [7]. TiO_2 nanotubes were synthesized by atomic layer deposition technology, providing a charge-conductive path and a low-weight scaffold for the SnO_2 flakes. Figure 4.2d illustrates the rate capability of the core–branch composite, which shows a specific capacity of 498 mAh g^{-1} at 3.2 A^{-1}. At the same

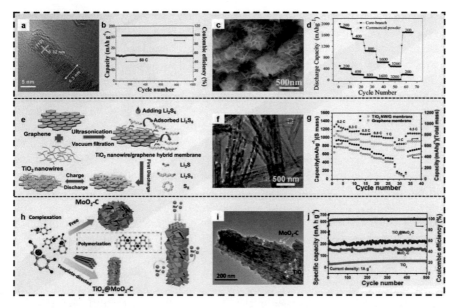

Fig. 4.2 **a** HRTEM images of a D–TiO$_2$ ultrafine nanorod. **b** The cycling property of D–TiO$_2$ at 50 C shown by the coulombic efficiency and Li$^+$ extraction capacity versus the cycle number [2]. (We shortened the original picture in the same proportion) Copyright 2015, Wiley-VCH. **c** SEM characterizations of the TiO$_2$/SnO$_2$ core-branch arrays. **d** The rate property of the TiO$_2$/SnO$_2$ core-branch architecture at various current densities versus the cycle number (the unit is mA g^{-1}) [7]. Copyright 2014, Elsevier. **e** Schematic of the fabrication process of a TiO$_2$ nanowire/graphene membrane and their interaction with the polysulfide lithium.) TEM image of the TiO$_2$ nanowire/graphene membrane. **g** The rate property of the TiO$_2$ nanowire/graphene membrane, and the pure graphene electrodes for Li/dissolved polysulfide cells at various current densities [9]. Copyright 2015, Elsevier. **h** A schematic illustration of the synthetic procedure of the TiO$_2$/MoO$_2$–C nanorod structure, and the ion transfer behaviors within this electrode matrix. **i** TEM image of the TiO$_2$/MoO$_2$–C construction. **j** The cycling properties of TiO$_2$/MoO$_2$–C, TiO$_2$ and MoO$_2$–C at 1 A g^{-1} [10]. Copyright 2017, Wiley-VCH

time, the coating of a thin mesoporous crystalline TMOs on graphitized carbon can provide a fast electron and ion transport pathway, which could provide high-powered lithium batteries. Therefore, flexible graphitized carbon covered by a mesoporous TiO$_2$ shell has been reported by a surfactant-templating pathway[8]. The flexible TiO$_2$/CNTs mesoporous nanocables deliver a high rate capacity of 210 mA h g^{-1} at 20 C and a remarkable cycling life with 210 mAh g^{-1} during 1000 cycles at 20 C. The synergistic coupling effect between the ultrathin mesoporous TiO$_2$ shells and the CNT cores, the highly crystalline mesoporous shells, and the available large pores and high specific area generate superior capabilities in lithium batteries.

Furthermore, a TiO$_2$ nanowires/graphene binary membrane has been successfully synthesized by high-power ultrasonication, vacuum filtration and drying steps [9]. The schematic illustration is shown in Fig. 4.2e. Moreover, the hybrid membrane (Fig. 4.2f) could be used as a free-standing current collector and could be easily peeled off. The hybrid membrane was mechanically robust and flexible, which could be used as an electrode for flexible devices. Additionally, the hybrid membrane for lithium-dissolved polysulfide batteries had a long cycling life and high capacity. The graphene membrane as a current collector reduces the internal resistance and immobilizes the dissolved lithium polysulfides. The TiO_2 nanowires not only exhibit an enhanced catalytic effect of the polysulfide oxidation and reduction but also have firm chemical binding with the Li-polysulfides, which promote fast redox reaction kinetics. This hybrid electrode shows a coulombic efficiency approaching 100%, with a capacity of 1053 mA h g^{-1} over 200 cycles at 0.2 C, a rate capacity of 850 mA h g^{-1} at 2 C and a remarkable specific capacity of 1327 mA h g^{-1}. The rate performance curve is shown in Fig. 4.2g. Significantly, achieving outstanding cycling stability and rate capability is the goal for sodium-ion batteries. Nevertheless, the large volume change and sluggish reaction kinetics mainly restrict sodium-ion batteries in the discharge and charge processes. As the schematic illustrated in Fig. 4.2h, Qiu and co-workers demonstrated the synthesis of TiO_2/MoO_2–C hierarchical construction [10]. The TiO_2 nanotube bunches were coated with a compound of the MoO_2 nanoparticles embedded in the C matrix, which demonstrates favorable electrochemical properties for sodium-ion batteries. The nanorod construction and MoO_2 nanoparticles of the TiO_2/MoO_2–C materials (as seen in Fig. 4.2i) can decrease the diffusion length of sodium ions and adapt the volume expansion. Figure 4.2j shows the cycling stability of the TiO_2/MoO_2–C, MoO_2–C and TiO_2 electrodes at 1 A g^{-1}. Additionally, this construction could achieve a cycling stability of up to 10,000 cycles at 10 A g^{-1} and shows a rate capacity of 76 mA h g^{-1} even at 20 A g^{-1}. In addition, an energy storage system with reliability, stringent safety and a high energy density is desirable for advanced next-generation energy storage devices. Shen et al. rationally designed an aqueous alkaline $Co_xNi_{2-x}S_2$//TiO_2 battery via the integration of two reversible electrodes associated with Li$^+$ insertion/extraction in the anode and hydroxyl ion insertion/extraction in the cathode [11]. The $Co_xNi_{2-x}S_2$//TiO_2 battery showed a remarkable cycling stability with a capacity retention of 75.2% after 1000 cycles, an outstanding energy density of 83.7 Wh kg^{-1} at 609 W kg^{-1} and a volumetric energy density of 21 Wh^{-1}. Due to the hierarchical electrode and unique battery configuration, alkaline $Co_xNi_{2-x}S_2$//TiO_2 batteries as large-scale energy storage devices are better than traditional commercial supercapacitors and thin-film batteries, demonstrating their potential as a promising battery technology.

4.2 Vanadium Oxides

As promising candidates as cathodes for LIBs, vanadium oxides, such as V_4O_9, V_2O_5, V_3O_7, V_6O_{13} and VO_2, etc., have been intensively studied because of the diverse vanadium oxidation states (e.g., V^{5+}, V^{2+}, V^{4+}, V^{3+}), wide availability, low cost and high specific capacities. Hierarchical LiV_3O_8 nanowire networks have been successfully synthesized by electrospinning using a polymer crosslinking approach. The cathode of LiV_3O_8 for lithium batteries gives a high rate capacity (202.8 mA h

g^{-1} at 2000 mA g^{-1}) and an excellent initial capacity (320.6 mA h g^{-1} at 100 mA g^{-1}) [12]. The decreasing charge transfer resistance, improved structural stability and large effective contact area resulted in the extraordinary performance. In general, this hierarchical LiV_3O_8 nanowire network is a hopeful cathode material for use in long-life and high-rate lithium batteries. To realize both the structural stability and the efficient transport of electrons or ions during the charge-discharge process, porous graphene/Fe_3O_4 nanowires (as seen in Fig. 4.3a, b) supported by amorphous VO_x matrices have been successfully synthesized by a phase separation process [13]. The porous graphene/Fe_3O_4/VO_x/hierarchical nanowires show a reversible specific capacity of 1146 mAh g^{-1} at 100 mA g^{-1} and 500 mAh g^{-1} at 5 A g^{-1}. Additionally, the Fe_3O_4/VO_x/graphene nanowires shows a capacity retention of approximately 99% after 100 cycles at 2 A g^{-1} (Fig. 4.3c). Notably, compared to V_2O_5, V_6O_{13} with mixed valence states is a less-researched vanadium oxide material, but it has exhibited better electrochemical property [14]. The double and single layers alternate in the mixed-valence V_6O_{13} crystal structure and share corners, offering more sites for Li intercalation. Yu et al. designed V_6O_{13} nanotextiles (Fig. 4.3d) at room temperature by a solution-based redox and self-assembly method. As illustrated in Fig. 4.3e, the V_6O_{13} nanotextiles delivered an energy density of 780 Wh kg^{-1}, which was higher than those of $LiCoO_2$ (540 Wh kg^{-1}), $LiMn_2O_4$ (500 Wh kg^{-1}), $LiNi_{0.5}Mn_{1.5}O_4$ (650 Wh kg^{-1}) and $LiFePO_4$ (500 Wh kg^{-1}). The rate performance of the V_6O_{13} nanotextile is shown in Fig. 4.3f. Furthermore, V_6O_{13} has an excellent theoretical specific capacity of 417 mAh g^{-1}.

In particular, surface engineering of the nanostructures plays a key role in optimizing the electrochemical performance of the battery electrode. In practice, VO_2 is a high-capacity but a less-stable unstable material that has been extensively used in the powder form as a lithium-ion battery cathode with ordinary performance. Chao et al. rationally designed and synthesized a binder-free cathode via a bottom-up growth process in which biface VO_2 arrays were directly grown on a graphene network for Na-ion battery and Li-ion battery cathodes (Fig. 4.3g) [15]. Furthermore, the VO_2 surfaces was coated with graphene quantum dots (GQDs) (Fig. 4.3h) as a surface sensitizer and for protection to further promote the electrochemical properties. The advanced electrodes show a remarkable capacity of 110 mAh g^{-1} at 18 A g^{-1} after 1500 cycles (shown in Fig. 4.3i) and a Na storage capacity of 306 mAh g^{-1} at 100 mA g^{-1}. Furthermore, $Na_{1.25}V_3O_8$ nanowires with a novel hierarchical zigzag structure (Fig. 4.3j) were rationally designed by a facile method [16]. Because the unique morphology provides better strain accommodation and increases the electrode-electrolyte contact were improved (Fig. 4.3k). This hierarchical zigzag material achieved a satisfactory cyclability with 0.0138% loss after 1000 cycles at 1 A g^{-1}, a high rate capability and a capacity of 172.5 mA h g^{-1} at 100 mA g^{-1} as a sodium-ion battery cathode. In addition, Fig. 4.3l shows the rate performance of zigzag $Na_{1.25}V_3O_8$ nanowires from 100 to 2000 mA g^{-1}. The morphology as well as the advantageous structure can synergistically facilitate the stability and kinetics, leading to a superior electrochemical performance.

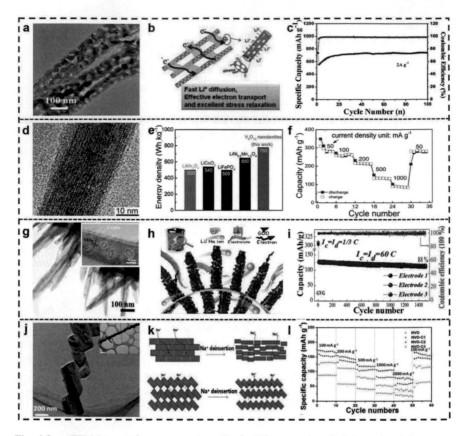

Fig. 4.3 a TEM image of porous graphene/Fe$_3$O$_4$/VO$_x$ nanowires. **b** Schematic of the porous nanowire and graphene composites with effective electron transport, fast Li$^+$ diffusion and outstanding stress relaxation during the process of Li$^+$ insertion and extraction. **c** Cyclic stability of porous graphene/Fe$_3$O$_4$/VO$_x$ nanowires at 2 A g^{-1} [13]. Copyright 2014, American Chemical Society. **d** HRTEM image of V$_6$O$_{13}$ nanotextiles. **e** The energy density comparison of this V$_6$O$_{13}$ nanotextile electrode and other conventional electrodes. **f** The rate capability of the V$_6$O$_{13}$ nanotextile electrode[14]. Copyright 2015, American Chemical Society. **g** TEM of VO$_2$/GQD nanoarray with its HRTEM pattern in inset. **h** Schematic of the VO$_2$/GQD electrode with bicontinuous Na$^+$/Li$^+$ and electron transfer channels. **i** Cycling property of three various electrodes after 1500 cycles at 60 C (We shorten the original picture in the same proportion) [15]. Copyright 2015, American Chemical Society. **j** TEM images of zigzag Na$_{1.25}$V$_3$O$_8$ hierarchical nanowires (NVO–C2). **k** Schematic of the electrochemical behavior for the simple nanowire construction and hierarchical zigzag nanowire construction of Na$_{1.25}$V$_3$O$_8$. The crystal structure of simple nanowire structure (upper) is not stable and they tend to aggregate. The topotactically synthesized zigzag nanowire architecture (lower) could provide excellently structural integrity and facile strain relaxation during the process of cycling. **l** Rate performance of zigzag Na$_{1.25}$V$_3$O$_8$ nanowire at various current rates [16]. Copyright 2015, Royal Society of Chemistry

4.3 Manganese Oxides

Manganese-based oxides are considered hopeful electrochemical materials for Li batteries because of their low cost, lower operating voltage, low toxicity, and high specific capacity [17]. Manganese oxides have multifarious phases include Mn_2O_3, MnO, MnO_2 and Mn_3O_4. 1D manganese oxide structures such as nanotubes and nanowires have been studied for many energy applications. Cai et al. designed a 1D yolk–shell nanorods of manganese oxide/carbon (Fig. 4.4b) by a two-step sol-gel coating strategy [18]. The precursors of tetraethoxysilane and dopamine were used to obtain a uniform silica layer and a uniform polymer coat by converting them into a hollow carbon and shell space, respectively. As illustrated in Fig. 4.4a, the interspace around the manganese oxide nanomaterials allows it to expand so the SEI layer is not ruptured during the lithiation/delithiation process and remains thin.

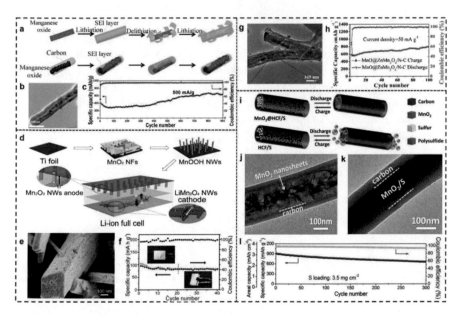

Fig. 4.4 **a** The illustration of lithiation and delithiation processes of bare manganese oxide nanorod (a) and yolk–shell manganese oxide/C nanorod. **b** TEM image of the yolk–shell manganese oxide/C nanorod. **c** The cycling property of yolk–shell manganese oxide/C nanorod electrode at 500 mA g^{-1} [18]. Copyright 2015, American Chemical Society. **d** Schematic illustration of the manufacturing process of the Mn_2O_3//$LiMn_2O_4$ nanowire for Li-ion full cell. **e** SEM image of Mn_2O_3 nanowires grown on the Ti foils. **f** Coulombic efficiency and cycling property of the flexible Mn_2O_3//$LiMn_2O_4$ nanowire full cell. The black arrows and optical images are shown in the insets [19]. Copyright 2014, American Chemical Society. **g** TEM image of MnO/$ZnMn_2O_4$/N–C yolk–shell nanorods. **h** Columbic efficiency and cycling performance of MnO//$ZnMn_2O_4$/N–C nanorods at 50 mA g^{-1} [20]. Copyright 2016, Wiley-VCH. **i** Illustration of the advantages of MnO_2/carbon nanofibers/S construction over hollow carbon nanofibers/S. **j** FESEM image of MnO_2/carbon nanofibers composite. **k** FESEM image of MnO_2/carbon nanofibers/S construction composite. **l** Cycling performance and Coulombic efficiency of MnO_2//carbon nanofibers/S at 0.5 C [21]. Copyright 2015, Wiley-VCH

This yolk–shell nanorod as an anode material of lithium batteries has an outstanding capacity of 634 mAh g^{-1} after 900 cycles at 500 mA g^{-1} (Fig. 4.4c) and exhibited a reversible lithium-storage capacity of 660 mA h g^{-1} at 100 mA g^{-1}. Because of the stability of the solid electrolyte interface, the electrochemical actuation of the manganese oxide yolk–shell nanostructure and the structural integrity, the capacity was enhanced during the long-term cycling process. The experimental results demonstrate that manganese oxide via the 1D yolk–shell nanostructure represents a beneficial method to realize enhanced electrochemical performance for practical applications. Additionally, homologous $LiMn_2O_4$ and Mn_2O_3 nanowires as cathodes and anodes, respectively, were designed and synthesized via a hydrothermal process to develop an all-nanowire, flexible lithium battery (Fig. 4.4d) [19]. This 1D nanostructure offers favorable charge transport, a short Li^+ diffusion path, and volume flexibility for lithium insertion and deintercalation, thus generating an outstanding cycling performance and rate capability. Figure 4.4e illustrates the SEM image of Mn_2O_3 nanowires. At the same time, the anode of the Mn_2O_3 nanowire shows an initial discharge capacity of 815.9 mA h g^{-1} at a current density of 100 mA g^{-1} and capacitance retention of 502.3 mA h g^{-1} after 100 cycles. The $LiMn_2O_4$ nanowire cathodes deliver a capacity of (approximately 94.7 mA h g^{-1}) at a current density of 100 mA g^{-1} and an excellent stability with 96% capacitance retention after 100 cycles. In addition, a flexible $LiMn_2O_4$ and Mn_2O_3 Li^+ full-cell was manufactured with high flexibility, an output voltage of more than 3 V, a specific capacity of 99 mA h g^{-1} and a low thickness of 0.3 mm. The coulombic efficiency and cycling properties of the nanowire full-cell are shown in Fig. 4.4f. Importantly, manganese oxides are promising electrode materials for lithium batteries; nonetheless, they generally display mediocre properties because of their intrinsic poor stability, high polarization and low ionic conductivity [20]. In the current study, the N-doped C coating on MnO yolk–shell nanorods coupled with $ZnMn_2O_4$ nanoparticles were synthesized by the carbonization of α–MnO_2/ZIF–8 precursors. The N–C MnO/$ZnMn_2O_4$ yolk–shell nanorod anodes (as seen in Fig. 4.4g) for lithium-ion batteries demonstrate a beneficial cycle performance with a capacity of 595 mAh g^{-1} after 200 cycles (at 1000 mAg^{-1}) and a capacity of 803 mAh g^{-1} after 100 cycles (at 50 mA g^{-1}). Figure 4.4h shows the cycling stability of the electrode at 50 mA g^{-1}. The favorable yolk–shell nanorod construction, nanoscale size and coating effect of N–C cause the optimized electrochemical performance.

Furthermore, lithium-sulfur batteries possess superior performance characteristics resulting from their superior theoretical energy density. The design principles of sulfur-based cathodes not only effectively restrict the polysulfides to alleviate their dissolution but also are highly conductive to strengthen the utilization of sulfur. Lou et al. successfully designed and synthesized a novel construction of MnO_2 nanosheets inserted into hollow carbon nanofibers (Fig. 4.4j) [21]. This nanocomposite structure not only efficiently reduces polysulfide dissolution but also facilitates ion and electron transfer in the process of the redox reactions (Fig. 4.4i). The composite electrodes have a sulfur content of approximately 71 wt% and areal sulfur mass loading of approximately 3.5 mg cm^{-2}. As illustrated in Fig. 4.4k, the sulfur was encapsulated within the carbon fibers. Significantly, the MnO_2/hollow carbon nanofibers/sulfur electrode yielded a remarkable specific capacity of approximately 1161 mAh g^{-1} at 0.05 C and maintained an outstanding cycle performance over 300 cycles at 0.5 C. Additionally, α–MnO$_2$ nanowires as a catalyst material for sodium-air cells at room temperature were designed and synthesized by a microwave-assisted method [22]. These cells were synthesized by a dry coating laser-induced forward transfer process without the use of organic solvents. The electrolyte shows an outstanding stability and electrochemical performance with a specific charge of 1215 Ah kg^{-1}. Direct laser-printing as an effective, green and fast pretreatment method has been widely used for sodium-air batteries. The laser-based printing process for sodium-air electrodes was verified as a significant technology for the preparation of air electrodes with high electrochemical performance. These results demonstrate that the poor magnesium-ion insertion and deintercalation capacities in manganese dioxide can be improved by introducing numerous water molecules into the electrolyte and synthesizing self-standing nanowires. The MnO$_2$ nanorod electrode cycled in a dry magnesium electrolyte after activation in a water-containing electrolyte exhibited an initial capacity of 120 mA h g^{-1} at 0.4 C and maintained 72% capacity retention after 100 cycles.

4.4 Iron Oxides

Owing to the non-toxicity, high theoretical capacity, improved safety and low cost, iron oxides, such as Fe$_3$O$_4$ and Fe$_2$O$_3$ have received extensive attention as anode materials for rechargeable lithium-ion batteries [22, 23]. Based on the principle of thermally driven contraction, NiFe$_2$O$_4$/Fe$_2$O$_3$ nanotubes (Fig. 4.5a) were fabricated using core–shell Fe$_2$Ni MIL–88/Fe MIL–88 as a precursor followed by an annealing step [24]. The experimental results show that the NiFe$_2$O$_4$/Fe$_2$O$_3$ nanotubes, which have diameters of 78 nm, are composed of numerous nanoparticles. The enhanced performance of the composite nanotube anode can be ascribed to the ameliorative design of the porous nanostructures and the typical synergetic effect of the binary functional components. The rate performance of the NiFe$_2$O$_4$/Fe$_2$O$_3$ nanotubes is shown in Fig. 4.5b; these hierarchical nanotubes deliver discharge capacities of 1392.9, 812.5, 586.7 mA h g^{-1} at 100, 500 and 1000 mA g^{-1}, respectively. Additionally, Shen et al. designed and synthesized porous Fe$_2$O$_3$ nanorods on CNT/graphene anode materials (Fig. 4.5c) for lithium-ion batteries [25]. The CNT/graphene foam (GF) network offers a high surface area, lightweight scaffold and high conductivity for the Fe$_2$O$_3$ nanorods. Figure 4.5d illustrates the rate performance of both the Fe$_2$O$_3$/GF and Fe$_2$O$_3$/CNT–GF electrodes in the range of 200 to 3000 mA g^{-1}. The remarkable properties could be ascribed to the electron transfer rendered by the CNT/graphene foam and the fast electrochemical reaction kinetics. Furthermore, the design ideas could be extended to other metal sulfides or oxides, providing a facile strategy for high-performance electrode materials.

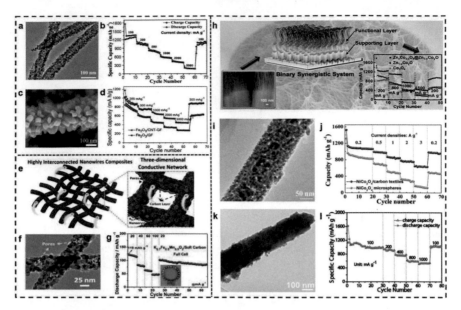

Fig. 4.5 a TEM image of the $NiFe_2O_4/Fe_2O_3$ nanotubes. **b** Rate performance of the $NiFe_2O_4/Fe_2O_3$ nanotube electrode [24]. Copyright 2014, Royal Society of Chemistry. **c** SEM image of Fe_2O_3 nanorods on carbon nanotubes. **d** Rate performance of Fe_2O_3/carbon nanotubes-graphene and Fe_2O_3/graphene electrode at various current densities [25]. Copyright 2014, Elsevier. **e** Schematic of the $K_{0.7}Fe_{0.5}Mn_{0.5}O_2$ nanowires with a large contact area and 3D continuous electron/K^+ transport pathways during the processes of K^+ insertion and extraction. **f** TEM image of $K_{0.7}Fe_{0.5}Mn_{0.5}O_2$ interconnected nanowires. **g** Rate capability of the $K_{0.7}Fe_{0.5}Mn_{0.5}O_2$ nanowires/soft carbon for K-ion full battery. Inset is the lighted LED driven by this full batteries [27]. Copyright 2016, American Chemical Society. **h** HRSEM image of $Zn_xCo_{3-x}O_4/Zn_{1-y}Co_yO$ precursor nanoarrays with fibrous-root construction (left). Illustration of fibrous-root $Zn_xCo_{3-x}O_4/Zn_{1-y}Co_yO$ nanoarrays (middle). Discharge capacity curves of Co_3O_4 nanoarrays, $Zn_{1-y}Co_yO$ nanoarrays and $Zn_xCo_{3-x}O_4/Zn_{1-y}Co_yO$ fibrous-root nanoarrays at different current densities (right)[31]. Copyright 2016, American Chemical Society. **i** TEM image of the $NiCo_2O_4$ naowires scratched down from the carbon textiles. **j** Rate performance of $NiCo_2O_4$ microsphere and $NiCo_2O_4$ nanowire arrays on carbon textiles at different current densities [35]. Copyright 2014, Wiley-VCH. **k** TEM image of the core–shell ZnO/ZnO quantum dots/C nanorod. **l** Rate capability of ZnO/ZnO quantum dots/C core–shell NRAs on CC anode at various current density [36]. Copyright 2015, Wiley-VCH

Significantly, iron oxides can be applied in rechargeable alkaline batteries with favorable capacity upon the completion of the redox reaction ($Fe^{3+} \leftrightharpoons Fe^0$). Nevertheless, their practical application was hindered by a weak structural stability. Liu et al. rationally designed and synthesized a Fe_3O_4-carbon binder-free nanorod array anode with a carbon shell protection strategy, which showed an excellent capacity of 7776.36 C cm^{-3} with 71.4% of the theoretical value, an outstanding rate performance and an enhanced cycling stability of more than 5000 cycles in alkaline electrolyte [26]. Moreover, a novel, flexible, solid-state alkaline rechargeable battery-supercapacitor hybrid device (with an approximate thickness of 360 μm) was assembled by pairing with capacitive carbon nanotubes. The remarkable energy density value approached and even exceeded that of thin-film batteries and is approximately several times that of traditionally commercial 5.5 V/100 mF supercapacitors. Significantly, the advanced hybrid device still maintains a favorable electrochemical performance in the case of elevated temperatures (up to 80°C), substantial bending and high mechanical pressure, demonstrating high environmental suitability. Furthermore, the Mn/Fe-based layered oxide connected nanowires were designed as a cathode for K-ion batteries, showing both good cycling stability and a high capacity (Fig. 4.5e) [27]. The experiment demonstrates that $K_{0.7}Fe_{0.5}Mn_{0.5}O_2$ interconnected nanowires (shown in Fig. 4.5f) could offer a 3D electron transport network, fast K^+ diffusion channels and a stable framework construction during the discharge process. These interconnected nanowires as electrode materials of K-ion batteries achieved an initial discharge capacity of 178 mAh g^{-1}. Furthermore, K-ion full batteries have been assembled by soft carbon and $K_{0.7}Fe_{0.5}Mn_{0.5}O_2$ nanowires, which show a capacity retention of 76% after 250 cycles. Figure 4.5g illustrates the rate capability of this nanowire/carbon composite for the K-ion full battery. The work above may speed up the research of high-powered K-ion intercalated layered electrodes.

4.5 Cobalt Oxides

Cobalt oxides is regarded as a promising alternative electrochemical materials due to its excellent electrochemical reactivity, high specific capacity and so on [28–30]. Yu et al. designed $Zn_xCo_{3-x}O_4/Zn_{1-y}Co_yO$ hybrid fibrous-root-like nanoarrays by a facile one-pot and successive-deposition process (Fig. 4.5h) [31]. This fibrous root nanostructure consisting of supportive and functional unique units can efficiently enhance energy and ion exchange. The fibrous-root-like composited structure was synthesized on a copper substrate as an integrated lithium-ion battery anode. Importantly, the stable $Zn_{1-y}Co_yO$ nanorods and ultrafine $Zn_xCo_{3-x}O_4$ nanowire units are synergistic in the supporting units of the multilevel array and, hence, produced superior rate performance with a current density of 500 mAg^{-1} and a discharge capacity of 804 mAh g^{-1} after 100 cycles. This composite electrode shows better performance than singular zinc-based or cobalt-based nanoarrays. These results demonstrate that an optimized electrode design is achieved by the hybrid synergistic nanoarray system. Furthermore, CoO nanowire clusters consisting of nanoparticles approximately 10 nm have been synthesized. The CoO nanowire clusters showed good reversible capacity at 1516.2 mA h g^{-1} at 1 C and 1330.5 mA h g^{-1} at 5 C [32]. This excellent rate capability and lithium-storage capacity are due to the typical hierarchical construction with fast Li^+ diffusion, satisfactory strain accommodation and a large electrolyte-electrode contact area. Notably, the experimental results show that the large additional capacity is derived from the higher-oxidation-state materials and pseudocapacitive charge. In addition, a longer cycle number can be achieved by coating SiO_2 shells on the hierarchical CoO nanowires. This strategy of coating SiO_2 on CoO to increase the cycle life could be extended to other transition metal oxides, for example CuO, MnO_x, Fe_2O_3, NiO and so on. Furthermore, Yu et al. successfully synthesized $(Co, Mn)_3O_4$ nanowires attached to a Ni foam hybrid structure without any conductive additives or binders [33]. This hybrid structure could be used as the air electrode for $Li-O_2$ batteries and showed a specific capacity of 3605 mA h g^{-1} at 0.05 mA cm^2. Continuous discharge and charge cycles with a maximum capacity of 500 mA h g^{-1} was obtained after fifty cycles. The specific $(Co, Mn)_3O_4$/Ni electrode provided a large capacity and high reversibility for the air batteries, increasing the void volume, which increase the accessibility to the reactants for the deposition of the discharge products at the surface of the nanowires. These results demonstrate that carbon-free $(Co, Mn)_3O_4$/Ni has great potential as air electrodes for $Li-O_2$ batteries. Due to the decreased performance levels observed from binder decomposition and the presence of carbon in tradition cathodes used in $Li-O_2$ batteries, using binder- and carbon-free cathodes are a hopeful solution. Lee et al. rationally designed vertical Co_3O_4 nanowire arrays attached to Ni foam [34]. This highly organized texture suppressed decomposition and showed a high catalytic activity with a reduced overvoltage. Furthermore, at a low discharge rate, a tapered brush morphology was observed, manifesting that the crystalline discharge results were concentrated on the surface of the cathode, causing deformation of the nanowire arrays.

4.6 Nickel Oxides

Binder-free $NiCo_2O_4$/carbon textiles (Fig. 4.5i) have been designed by a facile surfactant-assisted hydrothermal method and annealing treatment [35]. In this structure, the mesoporous $NiCo_2O_4$ nanowire arrays grown on the carbon textiles can be applied as electrodes for energy storage devices, such as supercapacitors and Li-ion batteries. Large, highly crystalline nanoparticles constituted the mesoporous $NiCo_2O_4$ nanowires. The abundance of mesopores alleviate the volume change during the charge/discharge process. Electrode architectures of large spaces between neighboring nanowires and plentiful mesoporous in the nanowires guaranteed fast electron transport via direct ion diffusion paths, and a facile connection to the growth substrate therefore guarantees that the nanowires participate in the ultrafast and efficient electrochemical reaction. Benefiting from the architectural features and intrinsic material characteristics, the binder-free $NiCo_2O_4$/carbon textiles show excellent rate capabilities (Fig. 4.5j), cycling stability and specific capacity/capacitance with an initial specific charge capacity of 1018 mA h g^{-1} with 77% coulombic efficiency. Additionally, the $NiCo_2O_4$/carbon textiles electrode exhibited the first discharge capacity of 1524 mA h g^{-1} and rapidly increased to 98% after a few cycles. Furthermore, the $NiCo_2O_4$ microspheres electrode shows an outstanding initial discharge capacity (1317 mA h g^{-1}). This textile electrode exhibits discharge capacities of 1357 mA h g^{-1} and 1012 mA h g^{-1} at 0.5 A g^{-1} for the first and second cycles, respectively. Moreover, the unique carbon-coated $NiCo_2O_4/SnO_2$ core–shell nanostructures were synthesized by a hydrothermal approach followed by a carbon coating technology

[37]. Both SnO_2 and $NiCo_2O_4$ are active materials for lithium storage, and the hybridization of SnO_2 and $NiCo_2O_4$ into integrated core–shell hetero-constructions could provide synergistic effects for effective lithium insertion and extraction. The $C/NiCo_2O_4/SnO_2$ hetero-constructions as anodes for lithium-ion batteries show an observably enhanced electrochemical performance. This promising anode delivers a remarkable reversible capacity of 654 mA h g^{-1} after 100 cycles at 100 mA g^{-1}. The outstanding lithium-storage properties could be attributed to the uniform carbon coating and characteristic core–shell hetero-constructions. Most importantly, complex mixed transition metal oxides (e.g., $NiCo_2O_4$) have been regarded as promising candidates for battery-like electrodes. In the present study, novel mesoporous $NiCo_2O_4$ hierarchical microtubes were fabricated by a self-templated method [38]. Due to their complicated constructions with super-thin subunits, the layered $NiCo_2O_4$ microtubes revealed favorable electrochemical properties in terms of their remarkable lifespan and high specific capacitance as a battery-type electrode for hybrid supercapacitors. This research provides a chance to bridge the electrochemical performance gap between supercapacitors and batteries. Importantly, a yolk–shell construction with specific chemical and physical properties is an attractive structure for electrochemical energy storage. For example, Lou et al. reported the yolk–shell Ni/Co hybrid oxide nanoprisms by thermal annealing in air, which were develop by a fast thermally driven contraction process [39]. The yolk–shell Ni–Co oxide electrodes yielded enhanced electrochemical performance for both lithium-ion batteries and hybrid supercapacitors. Furthermore, the $Ni_{0.37}Co$ oxide presents a remarkable specific capacitance of 1000 F g^{-1} at 10 A g^{-1} and an excellent stability with 98% capacity retention after 15,000 cycles.

4.7 Zinc Oxides

Zinc oxide materials have a high theoretical capacity in lithium-storage devices. Nevertheless, zinc oxide particles can detach from the current collector and become pulverized during the charge-discharge cycles due to large volume changes, resulting in its poor cycle performance. Sharma et al. fabricated ZnO particles in the hollow core of carbon-reduced graphene oxide by the electrostatic spinning approach. In the first step, a mat of core–shell nanostructured hybrid nanofibers was synthesized by a coaxial electrospinning method [40]. The ZnO nanoparticle core precursor and a carrier polymer and the shell of the reduced graphene oxide embedded in polyacrylonitrile ultimately comprise the core–shell nanostructure. Then, carbonization and calcination produce a stable anode material. First, it acts as a free-standing anode without a current collector and binder. The synergistic effect of the reduced graphene oxide, core–shell design and metal oxide lead to an enhanced capacity of 815 mA h g^{-1} at 50 mA g^{-1} and a capacity retention of 80% after 100 cycles. At the same time, Chen et al. synthesized ZnO/C porous hierarchical nanorods by a one-pot wet-chemical process followed by thermal calcination [41]. The experiment demonstrates that numerous nanograins comprise the ZnO/C porous nanorods, exhibiting an evidently hierarchical micro/nanostructure. The ZnO/C porous hierarchical nanorod anode exhibited a reversible capacity of 623.94 mA h g^{-1} at 1 C after 1500 cycles and showed long-term cycling stability and a high rate capability, which were attributed to the porous hierarchical structures of the ZnO/C nanorods and the excellent electronic conductivity of the modified PEDOT–PSS coating layer. In particular, metal organic frameworks are emerging organic/inorganic hybrid functional materials with a specific porous structure [42, 43]. The significant structural features of the MOFs are the high surface area and high porosity, which play an indispensable role in energy storage, [44, 45] gas separation, catalysis, [46] sensing, and drug delivery. An electrode material was fabricated by growing ZnO/ZnO quantum dots/C nanorod arrays on a conductive carbon cloth based on a scalable and facile ion exchange method (Fig. 4.5k) [36]. Due to the excellent electron collection efficiency, good structural stability and decreased lithium diffusion distance, this self-supported core-shell nanorod array yielded an excellent rate capability (Fig. 4.5l) of 1055 and 530 mA h g^{-1} at 100 and 1000 mA g^{-1}, respectively, and a cycling performance with 89% capacitance retention after 100 cycles at 500 mA g^{-1} (with a retention of 699 mA h g^{-1}). The ZnO/ZnO quantum dots/C nanorod electrodes provide a new direction and a new reference for ZnO-based nanomaterial in flexible electronic energy storage devices.

4.8   Other TMOs

Cupric oxide is a promising anode material for LIBs because of its high safety, high theoretical capacity, abundant sources and low cost, which are vital for energy storage devices [47]. Nonetheless, its practical application in energy storage devices is hindered by its morphological collapse and low conductivity, which can be ascribed to the volume expansion during lithium-ion intercalation/deintercalation. Hence, a porous and flexible CuO nanorod array was fabricated by a facile engraving method using copper foils in situ [48]. Furthermore, these foils could be used as the anode without any polymer binders or inactive conductive agents and further coating processes. The porous CuO array electrode for lithium-ion batteries exhibits an outstanding capacity of 640 mA h g^{-1} at 200 mA g^{-1} and a stable cycling ability.

To overcome the obvious degradation of MoO$_3$ rod-like nanomaterials in lithium batteries at high rates, surface-passivated MoO$_3$ nanorods were successfully synthesized via atomic layer deposition. The HfO$_2$-coated MoO$_3$ materials showed a higher specific capacity of approximately 68% after 50 cycles at 1500 mA g^{-1} than that of bare MoO$_3$ [49]. Additionally, HfO$_2$-coated MoO$_3$ delivered a specific capacity of 657 mAh g^{-1} after 50 cycles, and simultaneously, bare MoO$_3$ electrodes revealed 460 mAh g^{-1}. Furthermore, the HfO$_2$-coated MoO$_3$ materials were relatively better than bare MoO$_3$ electrodes because the nanoscale HfO$_2$ layer reduces the degradation of the MoO$_3$ rod-like construction. Furthermore, synergistic TiO$_2$–MoO$_3$ core–shell NW array anodes can be fabricated by a facile hydrothermal method followed by a controllable electrodeposition process. The nano–MoO$_3$ shell, which provides a large specific capacity and excellent electrical conductivity, played an important role in the fast charge transfer [50]. In the process of Li insertion/desertion, a very small volume change was noted, owing to the high electrochemical stability of the TiO_2 nanowire core, which improved the cycling stability of the MoO_3 shell. Meanwhile, there is a large amount electrodeposition of MoO_3. The array structure provides a 3D scaffold and contributes to an increased amount of electrodeposited MoO_3. The optimized TiO_2–MoO_3 hybrid anodes (mass ratio: ca. 1:1) have unique electrochemical attributes. They not only achieve a high gravimetric capacity (approximate at the theoretical value of the hybrid) but also have excellent cyclability (> 200 cycles). The TO–MO hybrid anode simultaneously exhibits a good rate capability (up to 2000 mA g^{-1}) and a high areal capacity of 3.986 mAh cm^{-2}.

Kabtamu et al. successfully synthesized high catalytic activity, stable and low-cost Nb-doped hexagonal WO_3 nanowires, which were employed as catalysts in an all-vanadium redox-flow battery, by a hydrothermal method [51]. This compound material can improve the catalytic activity of graphite felt electrodes using a redox-flow battery. The electrochemical impedance spectroscopy and cyclic voltammetry results showed that niobium-doped WO_3 nanowires with a niobium/tungsten atomic ratio of 0.03 showed excellent catalytic activities for VO_2^+/VO^{2+} among all the tested electrodes. Furthermore, the vanadium redox-flow single cell battery using the niobium-doped hexagonal tungsten trioxide nanowire (Nb/W = 0.03) catalyst showed an outstanding energy efficiency of 78.10% with 80 mA cm^2 in charge–discharge tests. This efficiency is higher than the vanadium redox-flow battery with untreated graphite felt (67.12%) or heat-treated graphite felt obtained through the traditional method (72.01%). Furthermore, the vanadium redox-flow battery with the Nb-doped hexagonal WO_3 nanowire catalyst in the stability test showed almost no decline after 30 cycles. The experiment demonstrates the excellent stability of the battery during the redox reaction of vanadium ions in acidic conditions.

References

1. Zheng M, Tang H, Li L, Hu Q, Zhang L, Xue H, Pang H (2018) Hierarchically nanostructured transition metal oxides for lithium-ion batteries. Adv Sci 5(3):1700592
2. Chen J, Song W, Hou H, Zhang Y, Jing M, Jia X, Ji X (2015) Ti^{3+} self-doped dark rutile TiO$_2$ ultrafine nanorods with durable high-rate capability for lithium-ion batteries. Adv Funct Mater 25(43):6793–6801. https://doi.org/10.1002/adfm.201502978
3. Aravindan V, Jinesh KB, Prabhakar RR, Kale VS, Madhavi S (2013) Atomic layer deposited (ALD) SnO$_2$ anodes with exceptional cycleability for Li-ion batteries. Nano Energy 2(5):720–725
4. Lin J, Peng Z, Xiang C, Ruan G, Yan Z, Natelson D, Tour JM (2013) Graphene nanoribbon and nanostructured SnO$_2$ composite anodes for lithium ion batteries. ACS Nano 7(7):6001–6006
5. Hu YY, Liu Z, Nam KW, Borkiewicz OJ, Cheng J, Hua X, Dunstan MT, Yu X, Wiaderek KM, Du LS (2013) Origin of additional capacities in metal oxide lithium-ion battery electrodes. Nat Mater 12(12):1130–1136
6. Zhang M, Li Y, Uchaker E, Candelaria S, Shen L, Wang T, Cao G (2013) Homogenous incorporation of SnO$_2$ nanoparticles in carbon cryogels via the thermal decomposition of stannous sulfate and their enhanced lithium-ion intercalation properties. Nano Energy 2(5):769–778

7. Zhu C, Xia X, Liu J, Fan Z, Chao D, Zhang H, Fan HJ (2014) TiO_2 nanotube@SnO_2 nanoflake core–branch arrays for lithium-ion battery anode. Nano Energy 4:105–112. https://doi.org/10. 1016/j.nanoen.2013.12.018

8. Liu Y, Elzatahry AA, Luo W, Lan K, Zhang P, Fan J, Wei Y, Wang C, Deng Y, Zheng G, Zhang F, Tang Y, Mai L, Zhao D (2016) Surfactant-templating strategy for ultrathin mesoporous TiO_2 coating on flexible graphitized carbon supports for high-performance lithium-ion battery. Nano Energy 25:80–90. https://doi.org/10.1016/j.nanoen.2016.04.028

9. Zhou G, Zhao Y, Zu C, Manthiram A (2015) Free-standing TiO_2 nanowire-embedded graphene hybrid membrane for advanced Li/dissolved polysulfide batteries. Nano Energy 12:240–249. https://doi.org/10.1016/j.nanoen.2014.12.029

10. Zhao C, Yu C, Zhang M, Huang H, Li S, Han X, Liu Z, Yang J, Xiao W, Liang J, Sun X, Qiu J (2017) Ultrafine MoO_2-Carbon microstructures enable ultralong-life power-type sodium ion storage by enhanced pseudocapacitance. Adv Energy Mater 7(15):1602880. https://doi.org/10. 1002/aenm.201602880

11. Liu J, Wang J, Ku Z, Wang H, Chen S, Zhang L, Lin J, Shen ZX (2016) Aqueous recharge-able alkaline $Co_xNi_{2-x}S_2/TiO_2$ battery. ACS Nano 10(1):1007–1016. https://doi.org/10.1021/ acsnano.5b06275

12. Ren W, Zheng Z, Luo Y, Chen W, Niu C, Zhao K, Yan M, Zhang L, Meng J, Mai L (2015) An electrospun hierarchical LiV_3O_8 nanowire-in-network for high-rate and long-life lithium batteries. J Mater Chem A 3(39):19850–19856. https://doi.org/10.1039/c5ta04643b

13. An Q, Lv F, Liu Q, Han C, Zhao K, Sheng J, Wei Q, Yan M, Mai L (2014) Amorphous vanadium oxide matrixes supporting hierarchical porous Fe_3O_4/graphene nanowires as a high-rate lithium storage anode. Nano Lett 14(11):6250–6256. https://doi.org/10.1021/nl5025694

14. Ding YL, Wen Y, Wu C, van Aken PA, Maier J, Yu Y (2015) 3D V_6O_{13} nanotextiles assembled from interconnected nanogrooves as cathode materials for high-energy lithium ion batteries. Nano Lett 15(2):1388–1394. https://doi.org/10.1021/nl504705z

15. Chao D, Zhu C, Xia X, Liu J, Zhang X, Wang J, Liang P, Lin J, Zhang H, Shen ZX, Fan HJ (2015) Graphene quantum dots coated VO_2 arrays for highly durable electrodes for Li and Na ion batteries. Nano Lett 15(1):565–573. https://doi.org/10.1021/nl504038s

16. Dong Y, Li S, Zhao K, Han C, Chen W, Wang B, Wang L, Xu B, Wei Q, Zhang L, Xu X, Mai L (2015) Hierarchical zigzag $Na_{1.25}V_3O_8$ nanowires with topotactically encoded superior performance for sodium-ion battery cathodes. Energ Environ Sci 8 (4):1267–1275. https://doi. org/10.1039/c5ee00036j

17. Chen L, Guo X, Lu W, Chen M, Li Q, Xue H, Pang H (2018) Manganese monoxide-based materials for advanced batteries. Coordin Chem Rev 368:13–34

18. Cai Z, Xu L, Yan M, Han C, He L, Hercule KM, Niu C, Yuan Z, Xu W, Qu L, Zhao K, Mai L (2015) Manganese oxide/carbon yolk–shell nanorod anodes for high capacity lithium batteries. Nano Lett 15(1):738–744. https://doi.org/10.1021/nl504427d

19. Wang Y, Wang Y, Jia D, Peng Z, Xia Y, Zheng G (2014) All-nanowire based Li-ion full cells using homologous Mn_2O_3 and $LiMn_2O_4$. Nano Lett 14(2):1080–1084. https://doi.org/10.1021/ nl4047834

20. Zhong M, Yang D, Xie C, Zhang Z, Zhou Z, Bu XH (2016) Yolk-Shell $MnO@ZnMn_2O_4$/N–C nanorods derived from alpha-MnO_2/ZIF–8 as anode materials for lithium ion batteries. Small 12(40):5564–5571. https://doi.org/10.1002/smll.201601959

21. Li Z, Zhang J, Lou XW (2015) Hollow carbon nanofibers filled with mno_2 nanosheets as efficient sulfur hosts for lithium-sulfur batteries. Angew Chem Int Edit 54(44):12886–12890. https://doi.org/10.1002/anie.201506972

22. Rosenberg S, Hintennach A (2015) In situ formation of α–MnO_2 nanowires as catalyst for sodium-air batteries. J Power Sources 274:1043–1048. https://doi.org/10.1016/j.jpowsour. 2014.10.187

23. Zhang L, Wu HB, Lou XW (2014) Iron-oxide-based advanced anode materials for lithium-ion batteries. Adv Energy Mater 4(4):1300958. https://doi.org/10.1002/aenm.201300958

24. Ma J, Guo X, Yan Y, Xue H, Pang H (2018) FeOx-based materials for electrochemical energy storage. Adv Sci 5(6):1700986

25. Huang G, Zhang F, Zhang L, Du X, Wang J, Wang L (2014) Hierarchical $NiFe_2O_4/Fe_2O_3$ nanotubes derived from metal organic frameworks for superior lithium ion battery anodes. J Mater Chem A 2(21):8048–8053. https://doi.org/10.1039/c4ta00200h

26. Chen M, Liu J, Chao D, Wang J, Yin J, Lin J, Fan H, Shen Z (2014) Porous α–Fe_2O_3 nanorods supported on carbon nanotubes-graphene foam as superior anode for lithium ion batteries. Nano Energy 9:364–372. https://doi.org/10.1016/j.nanoen.2014.08.011

27. Li R, Wang Y, Zhou C, Wang C, Ba X, Li Y, Huang X, Liu J (2015) Carbon-stabilized high-capacity ferroferric oxide nanorod array for flexible solid-state alkaline battery-supercapacitor hybrid device with high environmental suitability. Adv Funct Mater 25(33):5384–5394. https://doi.org/10.1002/adfm.201502265

28. Wang X, Xu X, Niu C, Meng J, Huang M, Liu X, Liu Z, Mai L (2017) Earth abundant Fe/Mn-based layered oxide interconnected nanowires for advanced k-ion full batteries. Nano Lett 17(1):544–550. https://doi.org/10.1021/acs.nanolett.6b04611

29. Li X, Wei J, Li Q, Zheng S, Xu Y, Du P, Chen C, Zhao J, Xue H, Xu Q, Pang H (2018) Nitrogen-doped cobalt oxide nanostructures derived from cobalt-alanine complexes for high-performance oxygen evolution reactions. Adv Funct Mater 28:1800886

30. Shi Y, Pan X, Li B, Zhao M, Pang H (2018) Co_3O_4 and its composites for high-performance Li-ion batteries. Chem Eng J 343:427–446

31. Li B, Gu P, Zhang G, Lu Y, Huang K, Xue H, Pang H (2017) Ultrathin nanosheet assembled Sn0.91Co0.19S2 nanocages with exposed (100) facets for high-performance lithium-ion batteries. Small 14(5):1702184

32. Yu J, Chen S, Hao W, Zhang S (2016) Fibrous-root-inspired design and lithium storage applications of a Co–Zn binary synergistic nanoarray system. ACS Nano 10(2):2500–2508. https://doi.org/10.1021/acsnano.5b07352

33. Cao K, Jiao L, Liu Y, Liu H, Wang Y, Yuan H (2015) Ultra-high capacity lithium-ion batteries with hierarchical CoO nanowire clusters as binder free electrodes. Adv Funct Mater 25(7):1082–1089. https://doi.org/10.1002/adfm.201403111

34. Lin X, Shang Y, Huang T, Yu A (2014) Carbon-free $(Co, Mn)_3O_4$ nanowires@Ni electrodes for lithium-oxygen batteries. Nanoscale 6(15):9043–9049. https://doi.org/10.1039/c4nr00292j

35. Lee H, Kim YJ, Lee DJ, Song J, Lee YM, Kim HT, Park JK (2014) Directly grown Co_3O_4 nanowire arrays on Ni-foam: structural effects of carbon-free and binder-free cathodes for lithium–oxygen batteries. J Mater Chem A 2(30):11891. https://doi.org/10.1039/c4ta01311e

36. Shen L, Che Q, Li H, Zhang X (2014) Mesoporous $NiCo_2O_4$ nanowire arrays grown on carbon textiles as binder-free flexible electrodes for energy storage. Adv Funct Mater 24(18):2630–2637. https://doi.org/10.1002/adfm.201303138

37. Zhang G, Hou S, Zhang H, Zeng W, Yan F, Li CC, Duan H (2015) High-performance and ultra-stable lithium-ion batteries based on MOF-derived ZnO@ZnO quantum dots/C core-shell nanorod arrays on a carbon cloth anode. Adv Mater 27(14):2400–2405. https://doi.org/10.1002/adma.201405222

38. Gao G, Wu HB, Ding S, Lou XW (2015) Preparation of carbon-coated $NiCo_2O_4$@SnO_2 hetero-nanostructures and their reversible lithium storage properties. Small 11(4):432–436. https://doi.org/10.1002/smll.201400152

39. Ma FX, Yu L, Xu CY, Lou XW (2016) Self-supported formation of hierarchical $NiCo_2O_4$ tetragonal microtubes with enhanced electrochemical properties. Energ Environ Sci 9(3):862–866. https://doi.org/10.1039/c5ee03772g

40. Yu L, Guan B, Xiao W, Lou XW (2015) Formation of yolk–shelled Ni–Co mixed oxide nanoprisms with enhanced electrochemical performance for hybrid supercapacitors and lithium ion batteries. Adv Energy Mater 5(21):1500981. https://doi.org/10.1002/aenm.201500981

41. Shilpa S, Basavaraja BM, Majumder SB, Sharma A (2015) Electrospun hollow glassy carbon–reduced graphene oxide nanofibers with encapsulated ZnO nanoparticles: a free standing anode for Li-ion batteries. J Mater Chem A 3(10):5344–5351. https://doi.org/10.1039/c4ta07220k

42. Xu G-L, Li Y, Ma T, Ren Y, Wang H-H, Wang L, Wen J, Miller D, Amine K, Chen Z (2015) PEDOT–PSS coated ZnO/C hierarchical porous nanorods as ultralong-life anode material for lithium ion batteries. Nano Energy 18:253–264. https://doi.org/10.1016/j.nanoen.2015.10.020

43. Wu R, Qian X, Zhou K, Wei J, Lou J, Ajayan PM (2014) Porous Spinel Zn(x)Co(3-x)O(4) hollow polyhedra templated for high-rate lithium-ion batteries. ACS Nano 8(6):6297–6303
44. Zhan WW, Kuang Q, Zhou JZ, Kong XJ, Xie ZX, Zheng LS (2013) Semiconductor@metal-organic framework core–shell heterostructures: a case of ZnO@ZIF–8 nanorods with selective photoelectrochemical response. J Am Chem Soc 135 (5):1926
45. Yang SJ, Nam S, Kim T, Im JH, Jung H, Kang JH, Wi S, Park B, Park CR (2013) Preparation and exceptional lithium anodic performance of porous carbon-coated ZnO quantum dots derived from a metal-organic framework. J Am Chem Soc 135(20):7394–7397
46. Cao X, Zheng B, Rui X, Shi W, Yan Q, Zhang H (2014) Metal oxide-coated three-dimensional graphene prepared by the use of metal-organic frameworks as precursors. Angew Chem Int Edit 126(5):1428
47. Li R, Hu J, Deng M, Wang H, Wang X, Hu Y, Jiang HL, Jiang J, Zhang Q, Xie Y (2014) Integration of an inorganic semiconductor with a metal-organic framework: a platform for enhanced gaseous photocatalytic reactions. Adv Mater 26(28):4783–4788
48. Zhang L, Li QY, Xue HG, Pang H (2018) Fabrication of Cu_2O-based materials for lithium-ion batteries. Chemsuschem 11(10):1581–1599
49. Yuan S, Huang XL, Ma DL, Wang HG, Meng FZ, Zhang XB (2014) Engraving copper foil to give large-scale binder-free porous CuO arrays for a high-performance sodium-ion battery anode. Adv Mater 26(14):2273–2279, 2284. https://doi.org/10.1002/adma.201304469
50. Ahmed B, Shahid M, Nagaraju DH, Anjum DH, Hedhili MN, Alshareef HN (2015) Surface passivation of MoO_3 nanorods by atomic layer deposition toward high rate durable Li ion battery anodes. ACS Appl Mater Inter 7(24):13154–13163. https://doi.org/10.1021/acsami.5b03395
51. Wang C, Wu L, Wang H, Zuo W, Li Y, Liu J (2015) Fabrication and shell optimization of synergistic TiO_2–MoO_3 core–shell nanowire array anode for high energy and power density lithium-ion batteries. Adv Funct Mater 25(23):3524–3533. https://doi.org/10.1002/adfm.201500634
52. Kabtamu DM, Chen JY, Chang YC, Wang CH (2016) Electrocatalytic activity of Nb-doped hexagonal WO_3 nanowire-modified graphite felt as a positive electrode for vanadium redox flow batteries. J Mater Chem A 4(29):11472–11480. https://doi.org/10.1039/c6ta03936g

Chapter 5
Summary and Perspectives

Abstract With popularization of portable electronics and the advancement of technology, portable, flexible, lightweight and wearable electronic devices have been greatly favored by the multitude of scientists and developers for their potential applications, such as artificial electronic skin, multidimensional stored energy devices and biocompatible electronic equipment. The adequately exploited of 1D/1D analogue TMO nanomaterials in new-type clean-energy equipment will create great opportunities to address the prospective challenges driven by the pressing environmental crisis and the increasing global energy demands. Nevertheless, many challenges still need to be addressed before real industrial applications. In this chapter, the challenges as well as perspective for the advancement of 1D-based TMO nanomaterials are discussed.

Keywords Transition metal oxides · One-dimensional architectures · One-dimensional analogue constructions · Batteries applications · Challenge · Perspective

1D-Based TMO nanomaterials are currently being intensively researched for use in the electrochemical energy storage fields due to their environmental friendliness, high theoretical capacity, moderate costs, admirable strain relaxation, large specific areas, correspondingly short electron or ion transportation path lengths and extraction/insertion characteristics. The results of both experimental studies and theoretical calculations have manifested that interface modification and structural control strategies could markedly enhance the stability, specific capacity, and rate performance of 1D TMOs. The keys to the aforementioned problems lie in designing high-level 1D nanostructures, including core-shell constructions, hollow structures, and multilevel structures, as well as hybrid architectures, which can facilitate higher surface areas and electron transport paths. Therefore, we summarize a set of high-efficiency and up-to-date synthetic routes, which include the Kirkendall effect, Ostwald ripening, heterogeneous contraction, electrospinning, photo-etching, template-assisted synthesis, liquid-phase preparation and vapor-based approaches.

Over the past five to ten years, numerous significant research results have been obtained for the manufacture of 1D nanoarchitectures with higher complexity in the aspects of interior construction, chemical composition, geometric morphology, and

© The Author(s), under exclusive license to Springer Nature Singapore Pte Ltd. 2020   71
H. Pang et al., *One-dimensional Transition Metal Oxides and Their Analogues for Batteries*, SpringerBriefs in Materials, https://doi.org/10.1007/978-981-15-5066-9_5

building blocks. The progress in fabricating these higher-order 1D structures has in turn accelerated their energy storage related applications. The increased complexity of 1D nanostructures could achieve more possibilities for modulating their electrochemical properties, which can increase their capability compared to their simpler configuration counterparts. Simultaneously, micro-sized frameworks based on 1D nanostructures (such as nanotubes, nanowires, nanorods and nanobelts) has afforded a way to integrate nanomaterials with a larger but more ordered and, thus, more manageable scope, into highly efficient catalysts, electronic equipment or bio- and chemo-sensors.

Here, we emphasized state-of-the-art 1D nanomaterials from their controllable synthesis and structural enhancement to their triumphant applications in various battery systems. We propose that the structural optimization and enhancement of 1D TMO nanomaterials, will overcome the limitations of many traditional electrode materials to realize a long cycle life, fast charge/discharge and high capacity. Furthermore, nanometer to macro integration problems of 1D materials are vital challenges in the use of nanostructures and calls for an unremitting progress towards novel material integrations and technological methods. Functional device fabrication with combinations of complex organized nanocomposites will be the target, and nanorods, nanobelts, nanotubes and nanowires are the nanostructures to achieve that. Along these lines, some prospective electrode design directions and challenge are discussed herein.

First and foremost, the construction of mixed transition metal oxides with a single-phase (such as, $NiCo_2O_4$, $NiMoO_4$ and $CoFe_2O_4$) is one of the best potential candidates for high-performance electrode materials. Their super-high electrochemical activity resulting from the multiple chemical constituents and their synergistic effects contribute to the outstanding specific capacity and capacitance, which are two to three times higher than those of the carbon or graphite-based electrode materials. Significantly, these mixed transition-metal oxides generally show higher electroconductivity than single transition metal oxides do because of the comparatively low activation energy required for electron transport between cations. Most importantly, the presence of multivalence cations in the mixed transition-metal oxide systems is beneficial to achieving satisfactory electrochemical action of the electrocatalysts towards the ORR for promising high-performance metal-O_2 batteries or fuel cells by providing chemisorption sites for the adsorption of oxygen. Additionally, the capacity could be observably heightened by controlling the morphology, surface area and pore properties. Manufacturing high-grade architectures (including core-shell, hollow and multilevel structures) is one of the most promising strategies for 1D TMO electrode materials. These highly advanced electrode configurations could provide more electrochemically active sites, higher specific areas and more effective contact with the bath solution, leading to greater charging and discharging capacities at high current densities. To further increase the volumetric and areal energy densities, we could make full use of holes within the electrode materials. However, this should be done in combination with other conditions. For instance, adequate space is still required for electrolyte wetting. Under certain conditions, these newly introduced active ingredients could function synchronously as "buffers" or "stabilizers" to strengthen the electrochemical cycling stability of the electrode. Most importantly, assembly from various strategies will be a fashionable trend for the preparation of some complicated 1D structures, where various reaction mechanism could be involved. Nevertheless, these synthetic routes are often limited to particular precursor substances. It is gratifying that apart from conventional inorganic templates (such as metallic materials/carbonates/oxides), organic functional materials (such as covalent organic frameworks), organic-inorganic hybrid materials (such as MOFs/metal alkoxides), and promising biological templates (such as M13 virus/bacillus subtilis/gram-positive bacteria/tobacco mosaic virus) are attracting widespread attention as functionalized precursors for high-level 1D nanostructures.

Finally, enhancing the electroconductivity of transition metal oxide nanomaterials to greatest extent by combining the materials with different conductive additives, for instance, forming composite structures to form metal oxides/carbon or metal oxides/conducting polymer. These composite nanomaterials could result in providing the advantages of both a high specific capacitance and high electrochemical conductivity. One important design is to utilize 3D current collectors to sustain the hybrid nanoarchitectures, which could supply the electrolyte to penetrate the electrode construction, which would achieve a large superficial area for ionic migration between the active material and the electrolyte, thus obtaining super-fast energy storage.